Video with Young People

Tony Dowmunt

Cartoons by Judy Farrar

Inter-Action Handbooks
CASSELL · LONDON

This Inter-Action Series
is dedicated to
E.J.B. Rose CBE
and Pamela Rose

First published in Great Britain in 1987 by
Inter-Action In-Print,
Royal Victoria Dock, London E16 1BT
and Cassell Publishers Limited,
Artillery House, Artillery Row,
London SW1P 1RT

This book was designed and produced by
Inter-Action Social Enterprise Trust,
Royal Victoria Dock, London E16 1BT.

Editor: *ED Berman MBE*
Sub-editor and Rewriting: *David Swarbrick*
Research: *Claire Sanders*
Design and Layout: *Myra Wareing*
Production Manager, Photography
and Technical Section: *Peter Mount*
Video in the Family: *David Powell*
Photography: *Roland Kemp*
Word-processing: *J Martha Powell*

Thanks and Acknowledgements
Jerome Chee-O-Tow, Tom Garrish,
Marianne Haddow, Vincent McGrath, Patricia Nanakwasi, Sam Powell,
Birmingham Film and Video Workshop
and Whitecross School.

Donors
British Petroleum Company plc,
Rank Xerox Trust,
Save & Prosper Educational Trust,
Tudor Trust and the Wates Foundation.

British Library Cataloguing in Publication Data see page 140

ISBN 0 304 31486 2

Printed and bound in Great Britain by Billings & Sons, Worcester

Contents

Preface

Video with Young People is a basic introduction to using video with young people in many different situations. Whether you are a parent, teacher, youth and community worker, play-leader or young person yourself, this handbook gives practical advice on how to set up worthwhile projects. The phrase "video worker" is used throughout the book to mean anyone in any of these roles working to help young people use the equipment constructively. It gets over the technicalities pretty quickly since all modern video systems are easy to use and give simple instructions in their operators manual. The emphasis in this book is on using video as a creative tool for exploration and discovery, creativity and documentation, individual skills building and co-operation in a team.

Take One! Project work on location in Leicester

Foreword

Is video really just a typewriter?

Inter-Action once ran a video project in a residential centre for people with cerebral palsy. We arranged for a wheelchair to be adapted by apprentices at BAC (British Aircraft Corporation). This enabled the equipment to be used by people who would not ordinarily be able to fine tune the controls. It was previously thought that their disability, spasticity, ruled them out of this type of activity. The opposite proved to be true; they used video to extend both themselves and their limited ability to communicate.

If such was the case with this severely handicapped group, it occurred to me that we needed to revise our ideas about who can use video and what are its real potentials and drawbacks.

The question of who can use it may seem like one to which the answer is "everyone". And, of course, in theory, everyone, except those with truly insurmountable handicaps, can indeed learn to use a VCR (video cassette-recorder) and camera.

As for its potential, again the answer may, in theory, be that video can be used to record anything. However, if we are not careful, video may simply become the visual equivalent of the typewriter. As it becomes increasingly less expensive and, thus, much more common to have such equipment, first in institutions and then at home, people might use it very much like a visual typewriter to record anything from how the family eats together to an English lesson in a classroom. The logical extension of this is boredom and the inevitable storing away of the equipment.

At least one part of the video process has become commonplace at home and in schools and youth clubs; video cassette-players to record programmes from broadcast TV. Most people with VCRs at home don't have cameras, however, and therefore don't make their own programmes. Furthermore, not everyone who has a camera will bother to learn to use video well; just as not everyone with a typewriter has become an expert typist. But most people will have tapped away at a typewriter at some time or another.

One of the dangers of this availability and ease of use is that people, young people especially, will begin to take video for granted. Instead of continuing their exploration with it and using it to develop social, intellectual and emotional abilities, video may simply fall into a rut. I have often heard teenagers say "Video!! Again?! But we've done that already!"

However, a machine that can produce quality television pictures – synchronized with sound for instantaneous playback and which can be used at a basic level after a few minutes training – has the broadest possible applications.

Specifically, any pictures or programmes recorded may simply be the starting point, the stimulus, for a far more important process – the involvement of people together in planning, analysis and expression. For young people this is especially important. Thus, anyone working with them should have the ability to use video as a basic tool in their skills tool kit. Families can become a video production team since making video programmes is best done in small teams.

Video can inform and help develop relationships, build up confidence, open a dialogue, be the excuse for a discussion, a study, survey or debate. It can motivate a disciplined approach to matters from planning to follow-through, producing documentaries or fictional material. It can thus stimulate the imagination and creative problem-solving abilities. Video can also be an objective outside eye, enabling us to focus on how we are seen by others and how we relate to others.

The process of enabling young people to express themselves through video, and the chain reaction that is engendered by making a programme and playing it back, is obviously more important than the technical quality of the programme itself. On the other hand, setting standards and learning more about the technical and editing aspects should be a further stage of development.

Enabling this involvement to happen is the job of parents, teachers, youth workers, play-leaders, specialists working with young people with difficulties, community workers, young people themselves and community artists. These latter are also known as "enablers" or "motivators" and in European jargon they are called "socio-cultural animateurs"! Every one of the above listed group-workers, whatever they may be called, should be trained in the use of video.

Video should be just one of their tools, however. The idea of a separate "witch-doctorhood" of video workers borders on insanity unless those workers are acting as a resource to all other community and social agencies in a given area. The main justification for this type of specialist is that few groups will ever need video full-time. Therefore, having a video resource in a centrally accessible agency might be a useful and cost-effective facility for many different groups in the community.

A parallel case in wasted resources are the thousands of duplicators, photo-copiers, cameras, projectors, vans etc. that are under-used because of the lack of co-operative sharing confederations and resource centres. Another justification for a specialist video resource centre is that the equipment needs constant repairing by somebody who knows what they're doing – a technician. Also, stand-by equipment will be needed so that the work can carry on when the technology fails. Much video equipment is locked up in the cupboards of schools or other institutions. This waste of capital can be solved by pooling the equipment centrally.

Simply having the ability and equipment to record and replay images will not help the cause of youth work or of individual or community development very much. The equipment, the process, the youth workers and other enablers must become out-going and involved.

That is why calling on the equipment when it's appropriate, rather than looking for things to do to fill up time, is the better approach to using the medium.

One of the greatest drawbacks to video is its very simplicity, coupled with the great expectations that have been trumpeted for it. Video is in danger of becoming the wonder-drug of community and youth work. Today's wonder-drug is tomorrow's pedestrian pill. Improvements in people's lives are hardly to be wrought by opiates.

Because of its immediate playback facility and its simplicity of operation, video is grasped at like a straw by drowning social, community and youth workers and activists. If an activity is going well, video might enhance it. If something is going badly, then the use of video will tart it up only for a while, without any lasting improvements; indeed, the effect may be negative.

It should be unnecessary to say, but video, like printing, is a medium of communication. It is not an end in itself. It can help explore and explain issues, initiate and develop relationships as work goes on to produce a programme and it can be a means to a wide variety of goals.

Video is seductive and flashy – instant accomplishment. The

problem is that people – young people especially – may use it, get used to it and discard it without ever exploring its real potential for their particular situation. If people are misled into expectations which cannot be fulfilled by video all the possible benefits will be destroyed.

Video in itself will not obtain housing, establish tenants' associations, make young people into responsible adults or change the world. It can, however, literally and figuratively, focus attention on the real issues. It forces people to get to grips with the nuts and bolts of problems because lofty words and principles are not sufficient to make an interesting programme that others will want to watch. Nothing is more boring to others than five minutes of unedited tape, however fascinating it is to the makers.

Video is still an intimate viewing medium for relatively few people at a time on a small screen, (used at home or in the classroom, youth club, etc.). In this sense, it's different from film shown on a large screen in a hall. Therefore, one of video's special characteristics is the playback. The delivery system of video tape to the audience – the playback – is an area which constantly needs attention. In most video work with young people the making and watching of the video tape often becomes the end in itself.

But sometimes young people, as well as adults, will want to reach a wider audience. They may have made their tape to inform, entertain or help others. The delivery system for such tapes has to be clearly thought through in advance – before making the programme. Who are we making this tape for? Where, when and how are we going to play it back?

A delivery system for video may mean something as simple as playing back in playgrounds or on a market stall with cables from the nearest power points. Or it may mean in-depth projects like some of those outlined in this handbook. You might join a tape exchange with others doing the same sort of work. Or you might set up a shop-front playback situation on the High Street, with an interview situation there as well. Or it may mean an elaborate self-contained media van such as Inter-Action's experiment – the Community Media Van. Whatever the choice of action, finding those who would not ordinarily feel able to be involved is critical. Video can be extremely useful for this purpose.

For the vast majority of people, television is their most important stimulus. Most people see and hear more "good" films, plays, paintings, documentaries, discussions, sculpture, music, etc. on the box than they ever will in the flesh. Young people are especially susceptible to this because of their impressionability. The very idea that young people can themselves produce the same sort of thing as on TV is a great confidence builder. These young people, who are often unwilling to take even a small step towards exploring themselves or their environment or working together, can get hooked on their ability to be seen and to put things on the TV with the aid of video cameras and recorders.

But it's what happens before and after the video process and product – how they plan and use the experience and the skills gained – that really counts. Young people particularly can be motivated by the product and can learn the following from participating in the process of making a tape with others: confidence, discipline, co-operation, planning, interviewing, analysis, and other social, personal and intellectual skills.

If this potential is not explored and encouraged by a worker through well thought out projects and activities, it won't fulfil the rest of its enormous promise. Video will simply become another typewriter – a sight and sound one.

ED Berman M.B.E.

Part 1: Video uses

1. Video: What is it?

Video is the name given to electronic machinery which records and plays back television pictures. A video camera converts pictures into electronic signals which are then recorded onto magnetic tape by a video cassette-recorder (VCR). Video tape is different from film as you cannot actually see the recorded image on the tape, and you do not have to wait for it to be developed – you can play the image back instantly through a television.

Video recording was invented in 1950 for use in broadcast television which, until then, had used only live transmissions or material shot on film. The original recording machinery was very bulky and complicated. By the late Sixties the electronics industry had seen a market for video equipment outside the broadcasting institutions.

The original recording machinery was very bulky and complicated

They began producing cameras and VCRs which were much smaller and lighter. These used half inch wide open reel tape. Hence the video cassette-recorder, or VCR, was developed as the most common video tape machine available. Throughout this book the term VCR will be used. It is very unlikely you will ever come across an open reel video tape-recorder nowadays, unless you start making broadcast quality tapes, and even those can now be made on high quality three quarter inch video cassettes. These open reel machines were sold mainly for industrial uses such as training, internal communication and security.

In the late Seventies two Japanese companies, Sony and JVC, made even smaller and simpler video tape-recorders using half inch tape in a cassette for home use. The domestic market (to differentiate from broadcast and industrial equipment) developed with such incredible rapidity that the cost of equipment actually came down, which made it very accessible.

Over the last ten years increasing numbers of people have been exploring the use of video in community campaigns, in education and youth work. Artists and other individuals, families and hobby groups are now using it. Because the equipment has become relatively cheap and simple to use, it's no longer the preserve of the captains of industry or those in the BBC or ITV. Ordinary people like ourselves can now use the powerful and popular medium of television through video cameras and video cassette-recorders.

Out and about with portable video equipment

2. Why use it with young people?

Firstly, because it's great fun. Most people are excited and involved by their first encounter with video.

Nearly everyone, especially young people, spends a great deal of time looking at the TV. It's estimated that people in the U.K. watch an average of over three hours of television a day. For the young it plays an important part in shaping the way they see themselves and the world. But like most of us, they are remote from the processes that go into making TV programmes, so the television has a lot of magic and glamour about it. Using video equipment gives them a chance to grab a little bit of that for themselves; they put themselves on TV and become stars for a few moments. While this is not to be condoned, it is inevitable. The positive spin-off is that in the process they come to a closer understanding of how TV is made. A more active involvement with the medium can help them to have a more questioning attitude towards what they see at home every day.

Familiarity with television means that young people are less reluctant to express themselves with video than they might be with, say, words or paint. The fact that equipment is easy to operate (it takes only a few minutes to learn the basics of camera operation) means that there is no need for young people to feel put off by having to learn complicated skills before they can use it. Because they can instantly play back material they've shot, they quickly get a sense of achievement.

All this is particularly valuable when working with less confident young people whose other experiences of learning may have led to them being labelled – or to feel themselves to be – failures.

Although video equipment is fairly expensive initially (a complete outfit costs about £1500), the running costs are fairly low as the magnetic tape can be re-recorded over a number of times. This is important as it removes the pressure to get it right first time. It also allows for experimentation and learning through mistakes. In fact, it's often true that the experience of using video is more important to young people than any product which may come out of it. It can be valuable on a number of levels more subtle than just that of making programmes.

An aid to discussion

At one level video can be a useful tool for the worker who wants to encourage young people to talk more. Merely showing a video tape can provide a focus for a topic which a group then might easily discuss under the guise of criticizing the tape.

Taking this a step further, a group discussion can itself be videotaped. This gives the participants a sense of its seriousness and importance. The topic under discussion can be developed by playing back parts of the tape selectively and talking about what has been said. Interviewing with video, a format already familiar from watching TV, provides an acceptable structure within which young people can explore each other's lives and opinions. Young people, indeed perhaps all ages, imitate these role models easily. Thus, there is little need to spend time cajoling or teaching the format.

Helping personal and group development

Being able to instantly play back what you have done is invaluable in all sorts of video work. It enables young people to evaluate themselves and each other by looking at the evidence on the tape instead of relying solely on the judgement of an adult authority. Young people in general prove to be very sensitive of what the camera reveals about themselves and their behaviour in front of it. They're also quick to notice, and criticize, wobbly or out-of-focus camerawork.

Video can be a very powerful tool in small group work. As one teenager, who had used video for a couple of years, put it: "You can only know yourself when you see yourself on telly... the way you act. It shows you what kind of person you are".

This opens it up to all sorts of uses (and abuses) and demands a great deal of sensitivity from adult workers. If, for instance, we ourselves are unwilling to be confronted by young people on aspects of our own behaviour which are revealed by video, then we surely have no right to put them through it. It seems unfair and manipulative to control the group from the safe area behind the camera while they're opening themselves up in front of it. It's also important to watch out for group members who use video to manipulate or put each other down.

Once everyone is alive to these problems, the group can use video to enact incidents from their everyday lives, or to rehearse interactions they're going to have with people outside the group. For example, a young person who has a lot of trouble at home can set up and tape an imaginary argument with their parents. By looking at the playback they might learn something about their own contribution to the problem, or what more appropriate responses they could develop. A group of unemployed teenagers who want to improve their job prospects could tape some mock interviews and see how they actually present themselves.

Rehearsing on camera — a group of children line up to record

Video is essentially a group activity. In any video project there are a number of roles to be played, both behind and in front of the camera. If the project is to be successful, the young people filling those roles need to be aware of what their responsibilities are and what the other people are doing. They need to develop a sense of themselves as a team, an interdependent group. They will pick this up quickly, as the more they organize themselves, the more coherent will be the result they get.

Video activity can make use of everyone's skills. There is no need for any member of the group to feel excluded. Those good at writing can work on ideas for the script, while the more visually oriented can paint scenery, make captions or use the camera. Extroverts gravitate naturally to roles in front of the camera, while a shy person would equally naturally seek to make as vital a contribution by operating it. However, apparently timid young people will often enjoy acting, especially if there's a chance to dress up as a completely different character.

Confronting the outside world

Finally, by using video young people can put themselves in a different relationship to the adult world, one in which they have more influence. They can go out and film the areas in which they live and comment on them. From behind the camera they can talk to and question adults who in other circumstances might intimidate them, and would certainly take them less seriously.

A lot of young people have found that making a video tape and showing it is one of the few ways open to them of bringing their point of view to an institution or group of adults who have power over them. Making programmes and showing them to parents, local councillors, school Heads, teachers, the local Police, Youth Officers or even a Council Committee are all situations where video has been successfully used by young people.

Part 2: Getting ready

3. Preparation

There are a number of areas to think about before starting to use video with young people. The crucial thing is to be very clear about why you are doing it. Video does present exciting possibilities, but by itself it's just a machine. It cannot substitute for the day-to-day work of building relationships with young people and it will not miraculously solve any of the problems you're having in that area.

Video versus other media

One primary question to answer is whether video is the appropriate medium for you to be using in the particular situation. It does have its limitations, especially in terms of the size and quality of the screen image, and these can make it the wrong tool for certain functions.

For example, the young people you're working with could want to publicize an event in their club or playground. Video would be far less suitable than printed posters or leaflets. It's difficult to show video to large numbers of people, especially if they are not particularly motivated to stop and watch it.

Photography, as some workers have found, can be a more satisfying activity in many situations for young people. They're better able to grasp the whole technical process, from taking the picture to enlarging the print, because it's more manual and more creatively involving. Although simple to operate, video works on very complicated electronic principles which are not easy to understand. Photography also allows young people the important reward of a tangible and readily portable product of their creative work: a photograph they have taken and processed themselves which they can take home with them.

If your group is interested in film-making they might get more out of working on Super-8 film than video. They would have the advantage of a better quality image which they could show on a large screen to more people than could cluster around a TV set. Film-making needs more care and discipline than video work, but if the group is sufficiently motivated this can be an advantage.

Once you have made the choice to use video there are a number of factors to consider. They're grouped below under five headings, under which are a number of questions which ought to be answered according to circumstance, or by anyone starting a video project.

The young people

Do they want to use video?
Because of the appeal of video it's easy to assume that the answer to this is "yes", but it's obviously important to be sensitive to their needs. Perhaps they want to do something more energetic, like playing sports. Or something more relaxed, like sitting around talking. Or something requiring individuals to work alone, like drawing, painting or writing.

What do they want to do with it?
If you have not met them before the video session, it's a good idea to spend half an hour or so chatting informally. Find out about their interests, what's going on in the club and local area, and what sort of a group they are. If they're serious, you may end up videotaping a discussion. A more light-hearted group may want to do a joke telling session.

How many young people are there?
There are definite limits to the number of people who can be involved at any one time. It takes, at most, only three people to operate video equipment, and to do anything more than play around with it takes a fair degree of concentration. These limits will obviously vary according to the number of adults around, what their organizing skills are and how highly motivated the young people may be. It's generally not a good idea to have more than ten young people with just one worker. Six tends to be an ideal number for one person to deal with.

What age are they?
Age must obviously be considered when you're thinking about what they want to do or can do. Most workers seem to feel that there is little point in using video with young people under seven or eight

years old. On the other hand, parents will be able to involve children of younger ages in a variety of tasks. How much do you expect the particular age group to learn about the equipment? What sort of ideas will they be interested in taping? Some workers have found that younger groups are more interested in fantasy, and that the older they are, the more willing they become about exploring aspects of everyday life.

There are definite limits to the number of people who can be involved at any one time

The adult workers

What do you want the young people to get out of using video?
Some adult workers may be very interested in making films and have expectations which may get in the way of appreciating what the group itself actually wants to do. Ask the members of the group what they want to do. If you want to point them in a particular direction, then give them just a few choices leading that way. It is easy to assume (and wrongly so) that there is nothing worthwhile going on in a session when the young people are using the equipment informally. It's more important to be sensitive and responsive to what is happening than to produce a finished tape every time.

How many of you are there?
It's best to have at least two adults working with a group, whatever the size, and to decide beforehand what roles you will each take

during the session. One of you, for example, can look after and teach the use of the equipment while the other organizes. Working on your own is fine if you're very energetic, know the young people well and are thoroughly familiar with the equipment.

How confident are you with the equipment?
It's important for you to feel in control of the equipment. The young people will be looking to you for answers if minor technical things start to go wrong. They'll become bored and frustrated if you don't know how to put things right. Your technical competence will make it easier for them to pick up the skills and your level of confidence will transmit itself. If this aspect worries you, try to get someone experienced to work with. However, the level of technical expertise required is not very great. Most of it can be acquired through a careful reading of the instructions which come with the equipment.

Equipment

Have you checked it?
Ensure in advance that you do actually have a complete set of working equipment for what you want to do. This is especially true if you're borrowing it. Leads and connectors are the eternal bugbear. Check that batteries are fully charged. Is there an operator's manual with the kit? Be sure to read it.

What do you do if it breaks down?
This applies particularly if you have planned a project which will take more than a single session. In this case, try to organize a back-up machine to borrow if the one you're using fails. Using a domestic set of video equipment with groups of young people demands far more from it than the designers ever intended. It may break down and can take a long time to be repaired.

An operator's manual is invaluable in dealing with minor problems with unfamiliar equipment. Most faults are visible as they're usually concerned with connectors, leads or switches which are simply not in the right position. Occasionally the picture breaks up due to dirty heads which should then be sprayed with head cleaner or by running a cleaning tape through the VCR. Often a more immediate but temporary solution is to move the tape forwards and backwards on VIDEO SEARCH to shift the dirt.

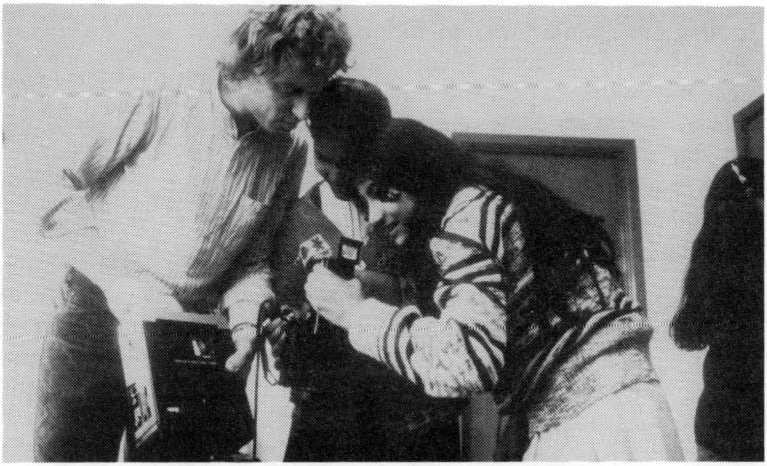

Running through the checks — a group check that the equipment is in full working order

Space

Where are you going to work?
It's generally best to start off working indoors. An enclosed space means fewer distractions, so the group can concentrate on learning how to operate the equipment and working out what they're going to do with it. If you're working in a club, try to use a separate room away from the main club area. At home, just ensure that distractions like the radio or TV are out of sight and hearing.

There are basic considerations about working indoors. You will need to be near a mains socket so that you don't continually have to run the battery down. The room needs to be well lit for the camera to function well; it should be as quiet as possible so that your soundtrack does not have too much extraneous noise on it.

Time

How long is the equipment available for?
Are you just doing the one session, a project lasting a few days or using it regularly with a weekly group? Will you have sufficient access to equipment to complete your plans? The answers to these questions will obviously affect the sort of work you're able to do.

Is there a "right time" to introduce video?
The answer to this depends very much on the nature of the group. As a general guide, it's probably not a good idea to start using video with a group that is for some reason unstable (if, for example, the group has started recently and the young people and workers do not yet know each other that well). It would be better to wait until relationships are better formed and a feeling of group identity is beginning to emerge.

How long can you expect to sustain their interest?
Again, the answer depends on the young people and their interests. If the group doesn't develop any particular fascination in video or the programme making process, it may be difficult to maintain energetic involvement in a project for more than three or four sessions.

Setting up a contract

It's important that ways are found of making the whole group – workers and young people – responsible for what is going on. One of the main points of using video is missed if the activity is entirely directed by adult workers. How successful the project is will largely depend on the young people and their interests. Work out what you want to do in advance. If you're going to do just a single session in which you want to do nothing more than give them a chance to play with the equipment and see themselves on the screen you only need a contract which outlines the basic points:

- An assurance from the group that they will take care of the equipment;
- That the camera will not be pointed at the sun or any other bright light;
- That they will see to it that everyone gets a turn to operate the camera.

It's important to do this before making the equipment available to the group, because once it's there the chances are they will be much too excited to listen.

Consensus – a group of young people negotiate a contract

If you're going to do more than one session, you will need to get the group to agree on rules and conditions for the entire project. Work out such things as how long you will take on it, when and where you will meet, who the members of the group are and whether new members will be allowed to join and on what terms.

Contracts can be either a verbal agreement or a more formal arrangement on paper which is signed by everyone involved.

Preparing the ground

There may be reasons for wanting to delay the introduction of the actual equipment to a group. You may have access to equipment for a few days only, in which case it could be a good idea to prepare them through doing other related activities first, so that they can make the most of the opportunity when it does arise. Or you may feel that video can have such a powerful effect that it's sensible to avoid introducing it too abruptly; you'll want to build up to it gradually through other activities. If, for example, some group members are not very articulate, they might find it less intimidating if their first experience of being recorded (and played back to the group) were on an ordinary sound cassette-recorder.

Some workers find great value in planning a project in considerable detail before introducing the equipment to the group. One reason for this is that you can then give them a better idea of the different aspects involved and the different roles they could perform on the project. For example, these might include designing and making costumes, scenery and captions, script writing, acting, interviewing, as well as operating the equipment. Without knowledge of these opportunities, young people tend to think that video work is only about camerawork and start competing with each other for the coveted role of camera operator.

Related activities

Photography is invaluable for basic training. It helps get the young people thinking visually. For example, it can demonstrate how easy it is to cut people's heads and feet off by framing incorrectly. Also, if you're working on a theme, the photographs could be incorporated in the video later on.

Framing the image on stills. Photography is one of the best starting points for people interested in video

Drama games and exercises help to lessen inhibitions and accustom young people to acting in front of each other.

Making Storyboards is a good way to start work on any video project. This is basically a representation on paper of the entire programme, with sketches to indicate how the images should look and related notes about the shot, dialogue, sound effects, etc. Prepare sheets of paper marked with a screen-shaped box and another for captions underneath, as the examples show:

> Have the young people sketch or paint their ideas for shots in this space and in the space below them ...

> ... write what action is happening in the shot, any dialogue between the characters and any other sound effects.

You could stick large sheets of marked out paper on a wall or in the middle of a circle so that they can work on the story as a group. Or they could make up stories individually on smaller sheets of paper.

Whichever way they do it, it introduces the idea that making video programmes consists of putting together sequences of pictures and sounds like a large jigsaw. It requires planning and organization.

Two young people are going into their local disco. The camera watches them as they happily discuss the evening ahead *(10 to 15 seconds)*.

They enter the disco and suddenly the lights start to flicker strangely *(10 to 25 seconds)*.

4. Introducing the equipment

You will need a basic working knowledge of portable video equipment to read this section. If you feel unsure turn to Appendix 2 ("The technical bit") for the basics.

From the beginning it's important to establish a good three-cornered relationship between you, the equipment and the young people you're working with. One of the biggest blocks to this is being over-protective towards the equipment.

It's very difficult to relax sufficiently to allow young people to experiment freely with such expensive equipment. Although they may often ask how much it costs, its value certainly won't stop them wanting to try it out. They tend to be much less inhibited than adults.

There is also a danger of the video equipment being damaged, so care does need to be taken about when to introduce it. Most of the time, though, young people respond well to the responsibility of using real equipment. The more responsibility they feel they have been given for it the more responsibly they tend to act.

There are a few fairly obvious ways of building up this trust. The moment young people see the equipment they'll want to touch it, pick it up and handle it. Don't stop them, but make sure you have set up in advance a system for who goes first and who goes next. If you show confidence in them, they'll have more confidence in themselves.

If you have to transport the equipment by car whenever you want to use it, you could make a point of entrusting the fetching and carrying to them. Make sure there is an inventory of all the equipment you take out. When they feel relaxed and confident with the equipment, it will increasingly be seen as their common property; consequently they will feel more responsible and involved in its use. In this way the equipment itself can bring the group together.

When young people are being particularly disruptive, often because they feel excluded or need to assert themselves, it's a good idea to let them handle equipment immediately rather than try to protect it from them. If they sense that you're putting unnecessary obstacles between them and the equipment, they'll feel justifiably frustrated and angry.

Starting off

A good way of showing young people what they can do with video is to set the equipment up in closed circuit. They can then all see on the monitor what the camera is pointed at. When they realize that they're on the screen there is always a strong reaction. The shy try to get out of the way of the camera; they hide their faces and giggle. The uninhibited get excited, peer into the lens or start performing. Everyone reacts powerfully and differently to the shock of seeing themselves.

It's important, particularly with younger people, to let this process take its natural course. It takes only a few minutes for the excitement to wear off. If you've worked out a contract in advance, you should soon be able to move on to working in a more structured way. At this stage try to impose only two rules:

- Never point the camera at the sun or any bright light (results in permanent damage);
- Everyone has the right to a turn looking through the camera.

It's preferable that people's first experience of video should be informal and playful. It breaks down many of the barriers between them and the equipment. It also helps to get rid of the feeling that making video programmes is an impossibly complicated skill which only a few people can ever hope to learn or use.

On the other hand, beginning in this way could be counter-productive for some young people. It might seem too easy. This could lead to their undervaluing what they're doing and losing interest. If you feel that the group you're working with would benefit from a more structured approach, then go straight into planning a project, or begin by teaching them in detail how to operate the equipment. This approach tends to be more appropriate for the older and more seriously motivated participants.

Another option, if the group is less inhibited, is to begin by playing a game which uses video. A good example is the Hand game. Get the group to divide themselves into pairs. One of each pair sits on a chair with the other sitting on their lap with their hands behind their backs, while the people underneath put their hands through their partners' arms underneath their armpits. The person underneath should not be seen by the camera, but their arms will appear to belong to the one on top – particularly if the shot is framed correctly.

You can then videotape an interview or conversation between two pairs, with the partners underneath gesticulating for their partners on top, who have the speaking roles. It helps to provide props, like a glass of water, or a handkerchief – anything that they can pass to the other pair or which demands an interaction between the hands and the rest of the body.

Playing this game can bring a group together in a very entertaining way. You don't often get a chance to scratch each other! It can also break down some of the barriers between adult workers and young people, provided everybody is willing to join in. When played back, the results are always hilarious.

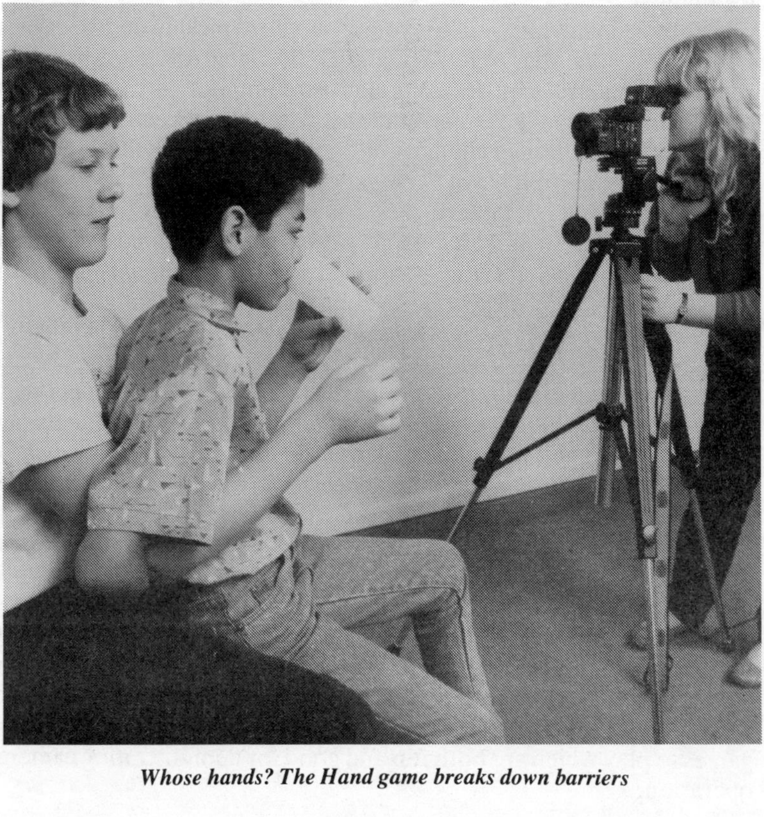

Whose hands? The Hand game breaks down barriers

Teaching the camera

After a few minutes of playing at seeing themselves on the screen, the young people will be ready to start learning how to operate the equipment. As a general rule, depending on the motivation of the group, it's not a good idea to try and teach them how to work all the equipment at once. For the first session it's normally sufficient for everyone to get some experience of the camera.

The equipment should still be in closed circuit so that everyone in the group (including the workers) can see what the camera operator is doing. The camera should always be on a tripod for its safety as well as for steadiness, especially when teaching. In this way it leaves both hands free to operate unfamiliar controls without any risk of dropping it.

The most important skill for young people to pick up at this stage is how to operate the ZOOM and FOCUS controls together; the routine of zooming in, focusing and zooming out to the shot they want. This will change every time the distance between the camera and what's being filmed changes. Without being shown this they tend to use the zoom on its own. They zoom into whatever they want to film and find that the picture goes out of focus. So they immediately zoom out again to a wider angle where the poor focus is not so noticeable, but where the picture shows more than is needed. Consequently, their subject is never brought properly into focus.

You can either show each young person this process individually, checking their progress on the monitor, or you can show one person and ask them to explain it to the person next to them; then get that person to explain it to the next, and so on round the group. Learning by teaching like this not only involves more than one young person at a time, it also puts the learning process more in the control of the group. Having to teach another reinforces understanding and team-building. This process also removes the adult from the role of knowing everything and avoids the boredom of repeating this elementary information.

Once young people do know how to use the camera, they will want to get on to make a programme. You could begin by working on an idea which someone in the group suggests. Or, if no-one comes up with anything they especially want to try out, there are a number of games to play which are both fun and good for demonstrating camera operation.

Games to play with the equipment

Mountain Climbing Tilt the camera at an angle, so that the floor in the picture seems to be tilted at an angle of 45 to 90 degrees. The young people can videotape themselves climbing up or slipping down simply by dragging their way across the floor. This can be developed further by having the camera operator tilt the camera first one way and then the other, while the group simultaneously hurl themselves across the room (and furniture) in time to it. On playback, the effect is like the passenger deck of a ship on a rough sea.

The Focus game This game is more suitable for younger children. Set the camera and monitor up so that they're divided by some sort of screen, and the only way to see what is happening on the camera side is by looking at the monitor. Get some of the young people to choose objects to place in front of the camera, and then the camera operator brings each item slowly into focus. On the other side of the screen, the rest of the group has to guess, as quickly as possible, what the objects on the monitor are.

The Mountain Climbing game – crawling on the floor can be climbing up the screen

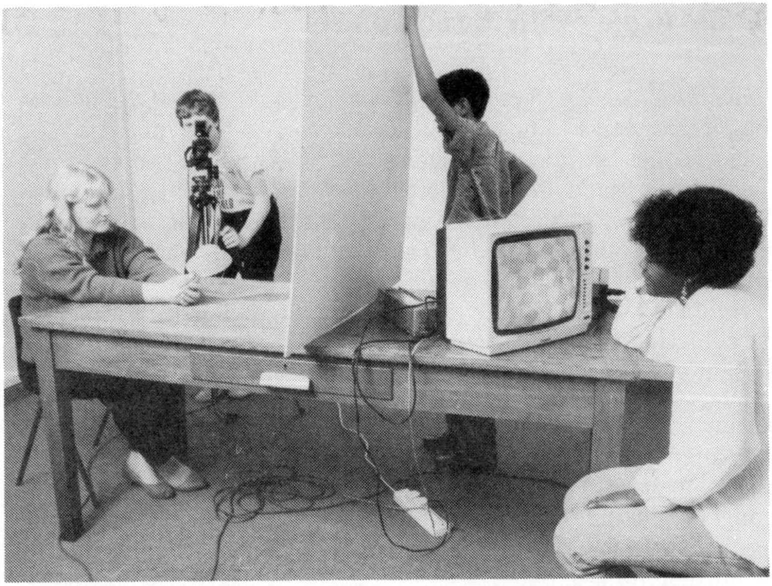

Defining the image. The Focus game tests the eye and the equipment

The Quincy Bent games Here are a series of three games taken from the book *Video and Kids*, devised by the American teacher, Quincy Bent. They provide an exciting way of helping young people to experience and understand what a camera does when it's pointed at a subject. They show that you cannot just record something without thinking about it. Making pictures is a conscious process, using the camera to think about the scene it is pointed at.

Game One Set the camera lens to wide angle and set the equipment in closed circuit, with the monitor placed so that everyone can see it. It's important that the camera is not moved. If you don't have a tripod set it on a table.

Ask a young person to move towards the camera until the screen is filled by their face. Now ask them to move away from the camera until their entire body fits into the screen.

Now ask a second person, then a third, a fourth, etc., to try to fit their whole bodies into the same picture. How many bodies can fit into the picture? If the room were longer would more fit in? If the room were shorter what would happen?

Game Two The idea is the same as in the first game, except that this time the people remain still and it's the camera which moves.

Ask one person to use the camera to fill the screen with someone's face. They may move the camera by hand or use the zoom control. Now ask them to pull back so that the subject's whole body fills the screen. How many bodies can be fitted into the screen?

Game Three Imagine that the screen is a kind of room, with a floor, ceiling and two walls represented by the bottom, top and sides of the frame.

Here, as in the first two games, the camera remains stationary and the subjects move. Ask one or more young people to walk into the room (i.e., into camera shot) and pretend to touch the screen floor. They can also pretend to lean their elbows on the screen floor. Then ask them to reach or jump up and touch the ceiling. Can they touch the left screen wall? They could also try leaning against it. Can they touch the right screen wall? Don't let them poke a hole in the pretend wall.

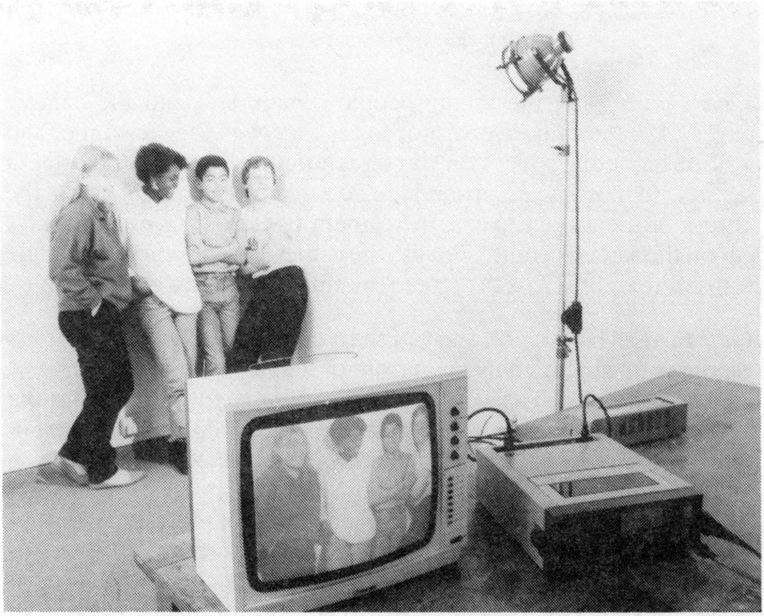

How large is a screen? Game Two — looking for its limits

Building barriers. Game Three shows the relationship between what you do and what you see

Game Four Now the young people remain still and the camera moves. Ask the camera operator to look through the viewfinder and position someone's head on the screen floor. Then ask the camera operator to move heads around the floor, from left to right across the bottom of the screen. Next, the camera operator places someone's feet on the screen ceiling. Finally, the camera operator moves people from left to right across the screen, bumping them into pretend walls.

Game Five This is a good way of sharpening up a camera operator's responses to what is going on in front of them.

Hide and Seek A: a young person must try to escape from the camera. They're not allowed to hide behind anything in the room but have to rely in quick movement, ducking and weaving. It's amazing how difficult it is to escape under these circumstances.

Hide and Seek B: this time the camera operator must try to prevent the subject from getting their picture on the screen. Because this game encourages lots of wild movement, it may be better to do it outside. It's a very good game to use if you're introducing video in an adventure playground.

Teaching the rest of the equipment

With any group that is going to run for more than one or two sessions, you could try doing yourself out of a job. In other words, try to get the young people to the stage where they can set up the equipment, record and play back without needing, or wanting, help or supervision. You won't arrive at that point by conventional instruction alone. It can be done through a mixture of demonstration and teaching over a number of sessions and by letting them pass on skills to each other wherever feasible.

Learning by practical experience will give them more confidence than being presented with a long series of rigid instructions. Allow the young people to experiment and to pick up the skills as they need them and at a pace that is comfortable to them. This is much more effective in the long run, both for skills retention, team-building and generating co-operation.

Setting up the equipment is quite straightforward; it's easily presented as a logical sequence. Familiarity and learning from a few mistakes will help them be aware of which switches to double-check and which leads go where. There are, however, a few points worth stressing to prevent potential problems later on.

Leads

Most breakdowns are due to faulty leads and connectors and most faults are due to mishandling. Stress the need for gentleness. Plugs are designed to fit sockets in a specific way; they can't be forced in upsidedown or off-centre.

Disconnect a plug by easing it out, never by pulling on the lead. Because plugs look so robust and substantial on the outside it's a good idea to illustrate the need for careful handling by taking one apart (a 5-pin DIN plug is a good choice) and showing just how delicate the insides are.

There is a technique for coiling cables which young people latch onto quite enthusiastically once they're shown it. The trick is to use the natural shape of the lead by giving it a half-twist with every coil and not to coil it too tightly. This is particularly important with Co-Ax cables which seem to suffer internal breakages rather easily.

Cassette care

Once it's discovered that the cassettes have a sprung flap, there's a temptation to try to pry them open. But the grease and acids from fingers damage tapes and mess up the magnetic heads on the VCR. Cassettes don't like being dropped or left in excessively hot or cold places. They also hate moisture, such as condensation and other wet or greasy things, such as tea, coffee, fruit and food.

There is also a non-liquid way to spoil a cassette. The image quality can be reduced, or even erased altogether, if it's left near a strong magnetic source like a loudspeaker, amplifier or TV monitor. Cassettes should be stored vertically to avoid tapes being crushed and buckled when other things are put on top of them.

Most breakdowns are due to faulty leads

The VCR

The most important component in a VCR is the recording head which spins at a very high speed. If the machine is moved physically while the head is spinning it will act like a gyroscope and try to stay in the same plane. At a minimum, this puts severe strain on the bearings which begin to get wobbly; this diminishes picture quality. Portable machines are designed to fare better when moved, but even they prefer to be as still and level as possible when recording.

Spilled liquids and food crumbs are obviously not wanted inside a VCR, but less obvious are hazards like rain, dust and puddles, especially with portables.

Making the most of sound

On many video tapes the most disappointing feature is the sound quality. It's important to remember that the audio quality is as important as the visual one. Both require care.

The first rule is always use a good separate microphone, not the one built into the camera. The microphone used on most video cameras is an omni-directional type (see Glossary) which picks up sound from all around. That includes passing traffic, the discussion in the next room, the noises of the camera trigger being switched on/off and handled and comments and conversation by the camera operator! Plugging in a separate microphone automatically cuts out the built-in one. When showing young people how to use microphones, you can expect three main problems:

- They have difficulty holding them still. Point out that any movement of their hands will result in loud thumping noises on the soundtrack. Try to use a stand of some kind wherever possible. For hand held work stress the need for the microphone to be firmly clasped so that the only moving part is the wrist; the fingers remain still;

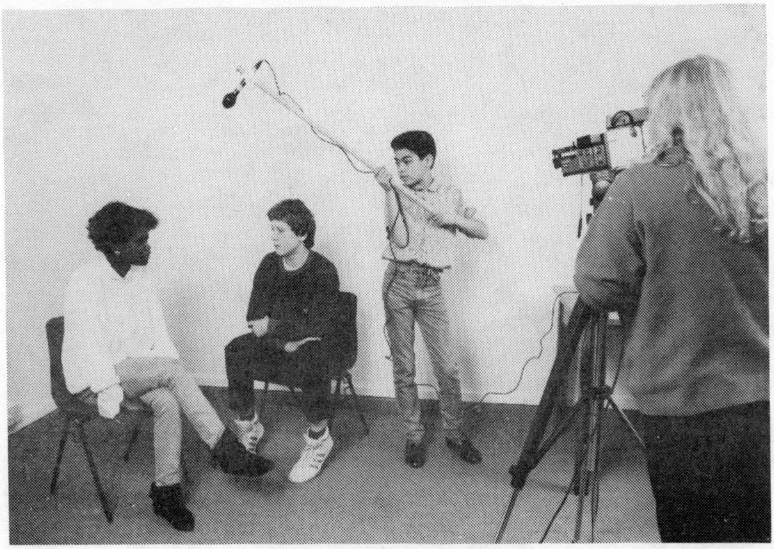

Sound – an added dimension

- They hold microphones too close to their mouths in imitation of rock stars on TV. The best results are obtained when the microphone is held about ten or twelve inches from the mouth;

- They whirl microphones around their heads, just like rock stars on TV (who can afford to keep buying new ones). This puts considerable strain on the internal connections and can result in a serious accident.

A separate microphone is always strongly preferable, even when you're shooting something like a drama where it has to be out of shot. The way to do this is by using a boom stand or a "fishpole". Fancy aluminium fishpoles are available, but you can improvise with a broomstick or something similar. With the microphone taped securely on to the end of a six foot pole, someone can hold it over the actors' heads or just out of camera shot at the bottom of the screen. The important thing is that the microphone points towards the sound source and is as close as possible to it. Fishpoles are also very useful for interviews, freeing the interviewer from worries about handling the microphone so that they can concentrate on obtaining good answers. The microphone boomstick is operated by another person who keeps it just out of camera shot.

Interviewing on location in the docklands

Interviews

When handling a microphone for the first time most people's reaction seems to be to want to interview someone. Perhaps we see so many interviews on television that it appears to be an easy and natural use for the medium. In reality, it's surprisingly difficult to keep going in an interesting way for more than a few minutes.

This is partly because the inexperienced tend to ask questions which can be answered by a simple "Yes" or "No". Consequently, they get demoralized by the lack of response and move on to another subject or give up altogether. The worker can help by suggesting ways to ask questions which demand a complete answer (e.g., "What do you think about...") and by pointing out that you can probe further by asking "Why?" even when faced with a "Yes" or "No".

It's also worth pointing out that people respond to an interviewer according to the amount of interest they perceive. If the interviewer appears to be bored, aggressive or embarrassed, it's unlikely that the interviewee will feel confident enough to open up and say what they think. It's been said that the art of interviewing is the art of listening and being interested in the answers you hear.

Video sound games

Commentary game Ask one of the group to give a continuous commentary on what is going on around them, such as a scene in a street or in a youth club. Ask the camera operator to illustrate the commentary by pointing the camera at whatever is being talked about and zooming in to close-ups of appropriate details.

You can also do this the other way around, by setting up the equipment so that the commentator can see only the monitor and thus has to follow the movement of the camera.

This exercise is a good way of illustrating the interdependence of the members of a video team, a point which is frequently lost in the initial excitement which overwhelms young people when they first use video. Without a degree of discipline, brought about naturally through such a game, it's easy to end up with soundtracks which have little to do with the images.

Sound dub game This game is best played with two sets of video gear (or half the participants will have nothing to do at any given time) and a story or a situation for them to act out in mime. Each group videotapes their mime. Then the completed, but silent, tapes are swapped. Using the AUDIO DUB button, they record their own soundtracks over the top of each other's images, following what they imagine to be the other group's story. Play both back to the whole group.

Finally, two more games from Quincy Bent:

Clapping and zooming One or more young people stand in front of the camera and clap slowly in rhythm. As their hands come together, the camera zooms, or physically moves, in. As they come apart, the camera moves out. When the young people become more adept at this they can elaborate on the theme by picking up the tempo, snapping fingers and so on.

Earthquake The game begins with the camera looking at an everyday scene of people sitting, standing and talking. Suddenly the camera starts shaking and tilting (not too violently) as someone makes appropriately thunderous noises into the microphone. The participants scream, fall and die dramatically as the earthquake strikes. The microphone can be a source of many off-screen effects. Blowing into it, for example, can sound like a rocket lifting off.

5. Working out ideas

Imagine your first video session with a group of young people. After they have all seen themselves on TV and are more or less familiar with operating thc camera and want to get on to doing something, what happens next?

Assuming that you did not start the session with a particular project or activity in mind, you will now have to make some choices. How structured do you think that session should be? How much should the young people follow their own noses or how much should they follow directions from you? Appropriate answers to these questions can come only from you, your knowledge of the group and of the situation you're working within.

Young people will usually come up with numerous ideas they want to try out, heavily influenced by their TV watching: spoof advertisements, police thrillers, interviews, horror and science fiction films. Occasionally they might want to enact incidents from their everyday lives.

A youth worker and children discuss which project to tackle

If you're working with the group for only one or two sessions and you do not have any long term objectives, it's a good idea to support them as they work on their own ideas. Suggest ways they could carry them out; point out what is impossible to do on video, write down plot lines and cast lists, and make sure that everyone in the group has a definite role – in front, or behind, the camera. This process is helped greatly if, before the start of the session, you have stuck large sheets of paper on the walls and have felt-tip pens to hand. As ideas are generated they can be written down for all to see. You can also use these later to make titles for the film you have worked out.

What can you do if the young people come up with no ideas at all? Some workers try to get round this by showing examples of work done by others. But the disadvantages of this are that if the tape is good they may feel intimidated. They will be concerned that they can't reach the same standard. If the tape is poor they have something to sneer at which might reduce their interest in using video themselves. Young people tend to be more motivated by, and involved in, their own use of video equipment than anyone else's.

The one circumstance in which it may be useful for one group to see another's tape, especially if they may know some of the people in it, is when they're frightened about appearing on video. Seeing an example of their friends getting away with it could give them a boost in confidence and enthusiasm.

Games and exercises

Perhaps a better way to get them thinking is to play some games. There are a couple of camera tricks you could show them which may stimulate their inventing stories.

The hat trick It's important that the camera stays absolutely still for this. Ask three young people to sit in a row, two of them wearing hats. Videotape a few seconds of that, tell them to stay very still, and pause the VCR. Another member of the group then swaps the hats around and the VCR is switched back on. Repeat the process a few times, changing, taking away and replacing different hats. Then play the tape back. Providing the camera has not been bumped, the hats will give a convincing impression of appearing and disappearing of their own accord. This is an enjoyable game to play for its own sake, but the technique also provides a good basis for stories about magicians, ghosts and other supernatural goings on.

The fight sequence Videotape a young person looking directly into the camera lens, acting and sounding violently angry and eventually throwing a punch at the lens. Immediately pause the VCR. Next, videotape another young person reeling backwards, falling over hurt, then picking themselves up and retaliating with a punch of their own at the lens. Pause the VCR, then shoot the first person reeling backwards, and so on. The result should be a remarkably convincing fight sequence which can lead the group into making up a story about the circumstances leading up to the fight, and/or what happened afterwards. Boys seem particularly keen on stories which climax in pitched battles. This exercise provides an exciting way of involving the camera in the action. It also demonstrates how subjective shots have a different emotional impact on the viewer when compared with a straight shot of two people fighting.

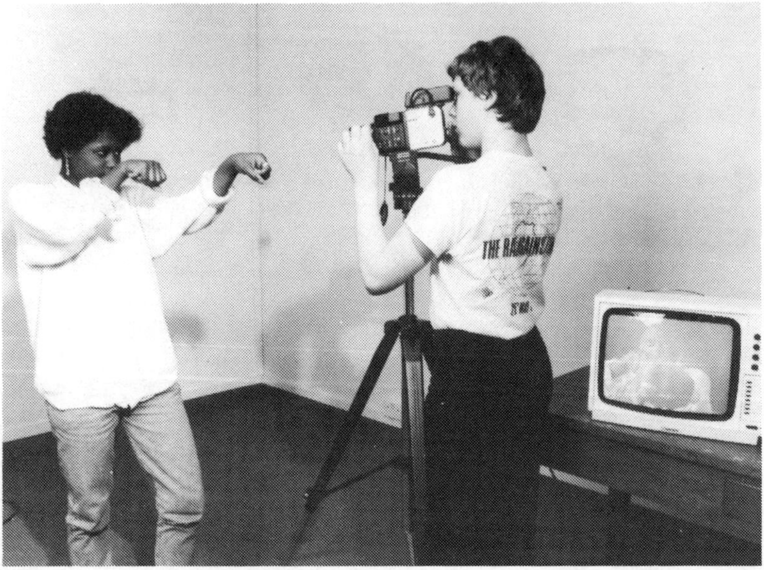

Fight sequences involve the camera too

Storytelling If you want to demonstrate how to tell a story more clearly in a few shots, you could ask them to videotape the following sequence devised by Quincy Bent:

Shot 1 Two young people are walking down the street on their way home from school. The camera watches them as they happily discuss the days events *(10 to 15 seconds)*.

Shot 2 Some robbers crouch behind dustbins waiting for their victims. They're anxious and whisper amongst themselves *(10 to 25 seconds)*.

Shot 3 The two victims look towards the camera, scream and attempt to run away *(5 seconds)*.

Shot 4 The robbers advance towards the camera looking really nasty (*6 seconds*).

Shot 5 A scuffle ensues involving both groups. First the robbers have the upper hand, but the victims finally beat them off *(20 seconds)*.

Shot 6 The camera watches as the two young people walk away from the camera, heading towards home. In the foreground are the unconscious bodies of the robbers. The camera fades out to black (by manually turning the aperture towards "C" or by operating the FADE control on your camera).

Developing stories

There are other games and techniques you can use to help the group start thinking of ideas for video tapes or to develop particular themes.

These games enable the whole group to participate in the planning of the video activity rather than leaving it to the more dominant or articulate members. A simple one, suitable for younger groups, is to play written Consequences, a game in which people build up stories by alternating the writing. Or you can start by sitting in a circle and telling a story, either by each adding a bit in turn or by giving each person a definite period of time (30 seconds or a minute). Alternatively, it can be done by passing a token around (such as a cup or ashtray) and whoever is holding it adds a bit to the story. You can videotape this exercise but it's not essential. Nor is it with the next game.

Clapping game This is a useful way of reducing inhibitions in a group and starting to work on a theme. The idea is to get each person to say something in turn, and to keep the pace up by having the rest of the group clap twice between turns. You could begin by going round the group saying something fairly trivial (like what everyone had for breakfast) and then move on to something closer to the theme which you want to develop. If, for example, the theme is unemployment with a group of teenagers, begin by having them say what jobs they wanted when they were kids. Or on romance, have them talk about their feelings for the first person who attracted them. This game can be used as a way of drawing out real life experiences to form the story. Anecdotes gained in this way can be developed by asking two of the group to go into the middle of the circle while others invent dialogue for them, one sentence at a time, going round the circle.

Photograph game Choose an image from an advert, a movie still, a magazine or newspaper. Get hold of some old clothes and props which roughly correspond to things appearing in the picture. Ask the people to reconstruct the image to resemble the original image osely as possible. Then ask them to make up a story which would lain the events leading up to the moment shown in the image or what happened after it. Videotape them acting out the story.

Scavenger game Choose a particular theme, such as differences in fashion, what people wear, or changes in the environment of a local shopping street. Ask the group to go out and videotape five different examples of the theme, or five different shots which explore the theme from different angles or tell a story about it.

Games about TV

The first thing many young people want to do when they start working with video is to make a programme which derives from TV. It may be a spoof TV advert or, if they're doing interviews, they invariably "sign off" like a TV reporter: "This is Jane Smith for *News at Ten*". Young people's fascination with broadcast television comes up again and again in their video work. Here are a few ways in which you can work with this fascination, using video to extend their understanding of television and how it works.

One point that can be made about TV (and the one which is most easy to demonstrate in practical work with video) is that it's not the open window on the world it pretends to be. Television does not present the whole truth or anything like it. It shows selected aspects of what is going on from a particular point of view.

You can begin to explain this by showing, for example, how altering the camera angle affects the way a viewer feels about a scene, or ways in which editing can influence how a scene is perceived.

Eisenstein's games Sergei Eisenstein invented many of the camera and editing techniques which are still universally used to affect a viewer's emotions. One example was to have his students shoot a piece of film looking down at a happy, gurgling baby. By editing this first with a shot of a laughing woman and then with an angry man, he completely alters the audience reaction to exactly the same shot, a child laughing with happiness, or one screaming in terror.

By using a low camera angle, looking up at a subject, the audience is led into seeing an authoritative and impressive figure. If exactly the same subject is shot from a high angle looking down on them, they seem ineffectual and unimpressive.

Setting up the angle – the camera can present a distinct point of view

You can see many techniques in use every evening on television, and their use really bears watching. In a drama they're an essential element in establishing and developing characterization and storyline. The same techniques used on the news, however, take on a totally different purpose.

The other games described below also show how decisions that programme makers take about how they deal with a subject inevitably mean that they show a particular point of view and not the whole objective truth.

Another commentary game This is a more structured version of the commentary game described earlier, designed to show how the commentary affects the way we see television pictures.

The group makes up a scene to act out (for example, an argument in the classroom or car accident) and then numbers off in pairs. When a number is called, one of that pair operates the camera and tries to follow the enacted scene, while the other watches it on monitor (turned away from the event itself), adding a commentary. When another number is called, those two return to enacting the incident

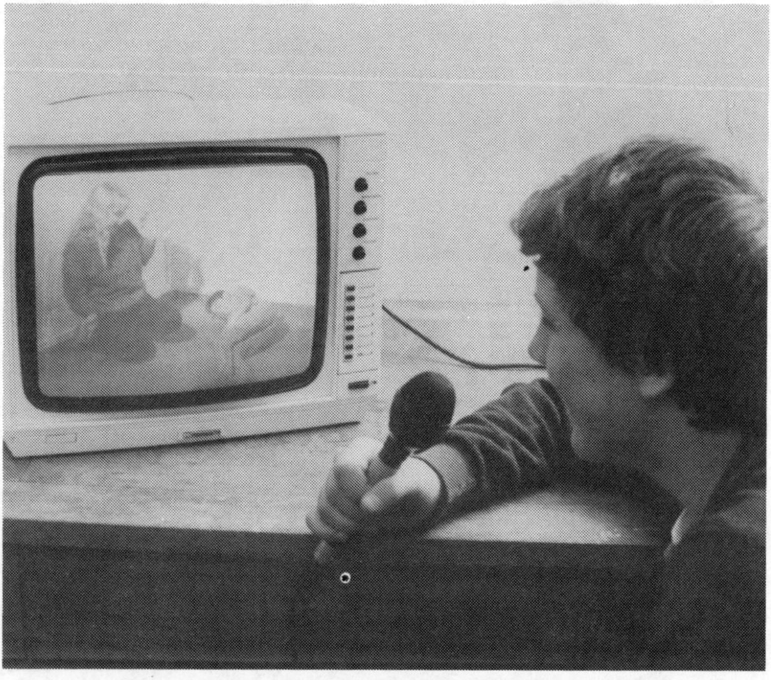

Filling in with words – young people take it in turns to add a commentary to a monitored scene

and are replaced by the next pair. In a quick sequence of number calls the whole group get to participate, both in the recording and scene playing.

When playing back the results you can point out and discuss the differences between each commentator's perspective and between the televised event and reality. Get the commentators to talk about the difference between commentating on an event they saw on monitor and being part of the event.

News story game Divide the group in two – a TV news crew and the news-makers. The latter have to invent a news story which could be about pickets at the factory gates, or victims and rescuers at an air disaster. You could use overturned tables and chairs to simulate the wreck. They have to work out the story in detail without letting the TV news crew hear what they're planning. The news crew have to arrive at the scene, investigate the story and videotape a short piece on it.

When you play it back the group can discuss how well they think the story they have worked out was represented on the screen, and whether it was an accurate and complete picture of what happened to them.

Chat show game Have the group divide themselves into roles: the chat show host, one or more guests, audience and camera operator. Videotape two versions of the show. In one, the host approves of the guest and asks friendly questions. In the second, the host makes no secret of disliking or disapproving of the guest. The audience participate by taking their cue from the host, applauding and laughing in the first version, booing and heckling in the second.

You could try having the guests as punks in the first version. The audience could pretend to be outraged sober citizens, and applauding the host's insulting questions. In the second, they could be punks themselves. When both versions are played back they demonstrate very effectively how the way a show is run can manipulate the viewer's sympathies.

Do your own news story A more ambitious way in which to use video to understand television better is to make your own version of a current news story. For example, after watching a TV report of a strike, have your group of young people interview some workers on picket duty. They might get a different idea of what the dispute was about than the one they saw in the original TV report.

This might be the sort of project you could go on to do after playing some of the games described above. It would require several sessions and a great deal of commitment from the young people to make it worthwhile.

Part 3: Projects

6. Resources for drama work

The basic resources for video work with young people are normally the video equipment itself and the space to work in. This space will usually be a small room or corner in a youth club, a classroom attached to a youth centre, or the sitting room or kitchen at home. Wherever it is, there's little chance that it will provide the perfect setting for drama ideas the young people develop unless, of course, they come up with something set in a school and you happen to be based in a classroom.

This section, therefore, sets out to show how to use unsuitable space to maximum effect. There are suggestions for a number of things you can bring to a session to help improve the look of your programme on tape.

One good thing about video is that you can do a lot with very little. Because the screen image is small and better suited to close-ups than wide panoramas, the atmosphere of a tape can be greatly enhanced by relatively simple devices such as a picture on the wall, simple costumes or suggestive scenery. It's not necessary for a set to look absolutely realistic to the naked eye in order for it to work on video.

Painted scenery backdrops can suggest a different place and need not be nearly as difficult or time consuming to make as you might think. Even very simple designs can be effective. One group created the impression of an airport by playing out the scene in front of a backdrop painted with a "Customs" sign and a roughly sketched aircraft tailplane. Paper was obtained for backdrops from a distributor of advertising hoarding posters. The posters themselves can make interesting backdrops or can be painted on the blank side. Newsprint also comes in wide rolls and can be used for this.

Small objects which you can use as props are most useful. Cups, saucers and a teapot add instant atmosphere to domestic scenes. It's also worth getting hold of an old telephone; the acting involved in one half of a phone conversation always seems to be enjoyable. Young people are very skilled at improvising from what is at hand if you can't get hold of the real thing. One group wanted to show a couple of bankrobbers driving away from the scene of the crime. They put two chairs in front of a wall and rocked around on them to suggest the movement of a car. The driver picked up a round waste paper basket and used it as the steering wheel.

Small objects which you can use as props are most useful

If you're working in the same place for more than a few weeks it would be a good idea to start a collection of props. You could ask the young people to bring in things they have lying around at home, such as old clothes, spectacles, make-up, old masks from joke shops and so on. It's particularly useful to have a large store of old clothes – perhaps the remnants of a jumble sale – as young people always enjoy dressing up and using clothes to create characters to play on video.

Sound is a most effective tool for creating atmosphere. Try to get access to a record player or cassette-recorder. Ask the group to bring in music they might like to use on the tape they will be making. You can then use music as part of the scenes you do, such as a record player in someone's room, or background music in a cafe. Music can also be added afterwards by using the AUDIO DUB button. Title sequences are much more effective with music behind them. Music with a fast rhythm also adds excitement to an action sequence – a chase, for example.

Perhaps the most useful tool there is for dramatic work with video is light. A group once wanted to do a scene involving an argument in a cinema. They sat in three or four rows of chairs with all the lights turned out except for a single powerful lamp pointed straight at their faces. They turned a TV on at low volume to get the sound effect. Fortunately there was a John Wayne western on at the time. While they were taping, a member of the group waved a card rapidly in front of the lamp to simulate the flickering of the cinema screen. They were all amazed at how convincing it looked when played back.

Inside every television is a child trying to get out

The flickering light trick can create a firelight effect by using yellow, orange, or red pieces of filter gel, or cellophane instead of card. In the film and television industry many backdrops are created simply with shadows made by cutting shapes out of sheets of card or aluminium. A large number of random holes create the dappled shadows of a forest glade and regular rows of holes make a science fiction or pop background. Recognizable shadows, such as the shape of a Gothic church window or trees, are actually very easy to make. A flashing blue light instantly says "Police". Red light adds authenticity to high-tech or science fiction sets. The list of possibilities is endless. Use a little imagination. Experiment yourself and never turn down the offer of a light source, whether an ex-theatre spot or a battered anglepoise. It's bound to come in useful.

A multitude of extras. Young people dress up to film a play in London's Kentish Town

7. Projects in the community

This section is about the sort of video work you can do outside a club, school or standard meeting place. It suggests some ways in which you can help young people to use video equipment as a way of exploring, learning about and even attempting to change aspects of their world.

Working outside

First, here is a Check List of technical things to remember if you intend to work away from your home or meeting place:

- Check the batteries' state of charge if you will not have access to mains electricity;
- Take a spare tape and a spare battery;
- Fit the microphone with a pop shield or some other wind shield – even a layer of foam rubber will help;
- Check that all the leads work and are undamaged;
- Check that the VCR carrying straps are properly fastened;
- Before you go out with a group, they should have had one or two training sessions with the equipment. It's much easier to show them how it works indoors. Make sure that those who will be carrying the VCR, microphone and camera are fully aware that they're all tied together by the cables. If they run off in different directions they will pull plugs out of sockets and wreck leads or connectors. So they need to be working effectively as a team before venturing out;
- Equip the VCR and camera with a "raincoat" if it looks as though it may be needed. A bin liner or plastic bag will work. Be careful where the VCR is put down in wet weather – even very shallow puddles are too deep for most VCRs to survive.

Check that the VCR carrying straps are properly fastened on

You don't need to have a formal project in mind to make use of video outside. A group may feel that, rather than staying indoors and working on a drama, they would prefer to go out and do some spontaneous interviewing on the streets.

Going out can also be a welcome relief or stimulus for groups who meet regularly indoors. One group met every week in a small room and occasionally used to go to the local fish and chip shop after the video session. There happened to be a television set there. So one evening they went along with the video equipment, shot some tapes between mouthfuls of chips and played it back on the TV. The other customers and staff enjoyed seeing themselves, and the group, who were normally quite introverted, enjoyed being able to use their video skills in public.

On another occasion, a small group from a holiday playscheme found a Greek Orthodox Church just off a busy high street. It turned out to be the cathedral for London and the South East. They met the Bishop who allowed them to tape while he gave them a conducted tour. He introduced them to the resident icon painter and to an old woman who claimed that the Virgin Mary had cured her of breast cancer. He then staged part of a service for the camera.

Those were relatively spontaneous examples of interacting with the community. Below are examples which require more pre-planning and forethought. In practice, you will find that projects of this sort develop from one to another. They also overlap the various programme categories which are listed for the sake of convenience.

Video in detached youth work

A number of detached youth workers have used video as a way of attracting and working with a group of young people they hadn't previously met. One worker contacted a group just by going on to their estate with portable video equipment. Initially drawn by the novelty of video and the fact that they were trusted to use the equipment, the group ended up making a tape about the lack of playspace in their area. They played this back to local Councillors and Council Officers who eventually agreed to give them a meeting room on the estate.

Effective use of the camera is open to everyone

As an aid to learning

Video provides young people with an excellent tool for researching and documenting aspects of their environment. A few years ago a group from a youth club in North London did a project on housing in their area. The following extract is from a report by the video worker who helped them:

> They're all a great deal more knowledgeable now about housing in Islington, and how the situation affects them, than they were when the project started. They have all gained considerably in confidence from interviewing so-called experts, who they would not have come into contact with in the past. A great deal of this confidence has also resulted from the technical skills learned in operating the equipment, and a fairly thorough political and social analysis has resulted from their researches into the paradoxes of homelessness in the Welfare State. The group is now quite stable and intends to make tapes on this and other topics to take into schools, youth clubs and tenants' meetings.

Video enables young people to approach subjects in a particularly vivid way. On a holiday project a group of teenagers made a video called *Work*. They interviewed a range of people at a local company, from the managing director to the apprentices on the shop floor.

Understanding video — a route to learning technical and social skills

When they played back the tapes, as well as discovering that the workers had very different attitudes in their commitment to their work and their feelings about unions, for instance, the young people could also match these attitudes to the way the workers came over as characters on the screen. Video also helped the workers to understand how they related as people to their work.

As in most educational work, the closer these sorts of projects are to the experience of the group you're working with, the more they will tend to get out of it. For example, video has often been used with school leavers and unemployed teenagers. With it they can investigate the types of jobs open to them by going round to different workplaces, videotaping what goes on and interviewing employees and employers.

As a tool in community work

Here are a few examples of how young people can use video to contribute to a process of community organization.

Some tenants on a London housing estate had the idea of starting a City Farm on a patch of open space which was going to waste. A worker gathered together a group of local young people and went with them to videotape the animals and activities at a well established City Farm in another area. They made a short tape by inter-cutting this material with shots of their own estate and played it back to the tenants to generate interest in the idea.

A project does not necessarily have to be tied to a specific issue to benefit the community in which it's made. If you have a local festival in your area, it's a good idea to involve a group of young people in videotaping the day's events. As well as having a record, the group can also show it the same evening in the local pub or community centre. Playbacks of this sort of material tend to be very popular as people recognize each other on the screen; they remember and celebrate what has happened during the day.

An area of South London contained a number of old housing estates and one new estate which had been filled with specially selected "respectable" tenants. This special treatment had produced a great deal of tension and division. It was aggravated by the fact that the whole area was supposed to be served by a new community centre being built right in the heart of the new estate. A community worker in the area was in contact with a theatre group and a video worker. Together they decided to run a playscheme with the underlying intention of bringing the community together.

They made a tape with a large group of local children about a monster on the loose, threatening the community – a theme borrowed from *King Kong*. The tape was shot on locations all over the area and played back many times in a local pub and on a stall at a neighbourhood festival. It became very popular with young people and adults in the area. Unfortunately, the community worker left the area shortly after the tape was completed, so any more accurate assessment of its affect on the life of the community was rather difficult.

It's worth mentioning that real communities tend to be more fragmented than the singular term "community" implies. Just because people live in the same locality does not mean that they share, or can be encouraged to share, the same interests. They're equally likely to be divided in ways that are important to recognize – by age, sex, class, or race, for example. Much careful thought needs to go into the planning of work in the community. For one thing, young people might not be that keen on getting involved in the lives of adults around them. They're more likely to get most out of working on something that closely reflects their own interests and their own place in the community.

Bridging the gap — video can help bring a community together

Presenting views

There are numerous institutions in our society which provide services for young people yet none are controlled by the young people themselves. Few make any formal attempt to find out how young people feel about what is provided for them. Video, however, can help young people gain more influence in this situation.

A video worker describes how a group on a playscheme made a tape:

> They interviewed both adults and children on their estate to poll opinion. They toured possible sites and compared visually the activities of a nearby adventure playground to their own sterile tarmac area. This video tape was shown back several times – once to the others at the playscheme, amid gleeful shouts at their first television appearance; next, to the annual general meeting of the Residents Association; and finally to relevant Council Committee Members. All were impressed by the product and could scarcely believe that children did it. Most importantly, video was the only medium that could command the attention of the Council Members and bring the children's viewpoint to them.

It seems that witnessing young people talking about themselves and their needs on a TV screen enables adults to take them more seriously. This was certainly the experience of a group of female youth workers who made a tape about working with girls. It included a number of interviews with girls complaining about their position in youth clubs and talking about how they enjoyed meeting and doing activities which did not include boys.

They played it back at a number of youth workers' meetings and were struck by how readily the workers reconsidered their attitudes to girls after looking at the tape. They also showed it to their Borough Youth Committee, who immediately decided to allocate money and resources to look into how girls' needs could best be served within their provision. Of course, it's only rarely that a tape of this sort has such an immediate effect.

Another group of girls chose to do a video option in their final year at a South London school. For the first three sessions they just talked. It emerged that they all hated school and had been truanting regularly. Three of them were from single-parent families and shared the experience of this being interpreted by Education Welfare Officers (EWOs) as the reason for their truanting. The group went on to videotape role-plays; some of these were on experiences in schools while others were on confrontations with EWOs in their homes.

The group decided to make a longer, better planned tape. The first scene showed them skipping school. It then followed them through a typical truant's day – riding a bus to the end of the route and all the way back again and hanging around in the local shopping centre. The tape then showed a confrontation between an EWO and one of the truant's parents. The EWO tried to blame the truanting on relationship difficulties at home.

This tape was played back on a number of occasions, first at the school and later to a large group of local Education Welfare Officers. This led to a lively discussion. Some of the EWOs were sympathetic while others felt that it was a dangerous film because it promoted prejudice against, and a misunderstanding of, their role. The video workers who helped the girls make the tape pointed out that its value was not that it was objective or balanced, if such a thing were possible, but that it was an accurate reflection of the girls' point of view, and, as such, should be watched and listened to.

8. Video in schools

You may decide to use video with young people in a school. Whether you're already a teacher working within a school or a project worker coming in from the outside, there are certain common problems that you have to face.

Despite the push now being given to Media Studies, most schools tend to have facilities for playing back programmes rather than creating them. On top of this, the school curriculum doesn't really allow for long periods of concentrated work. Even if you were able to get a series of double periods there is always the danger that young people will feel that they are spending most of the time assembling and dismantling equipment. So how can you get around these problems?

One of the best ways of answering this question is to look at the work of people who have done successful video projects in schools.

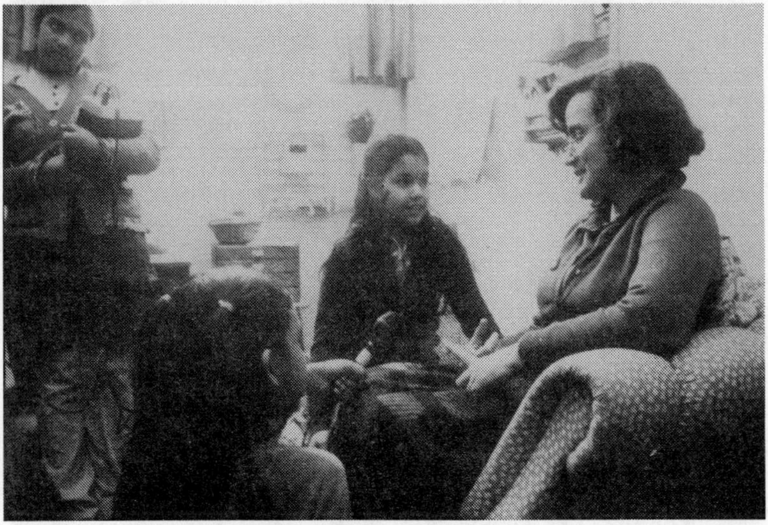

A group of children interview in school

Help the Aged project

Run by Bryan Rendell and Keith Childs at Whitecross School in Dean, this oral history project was initially a response to a Help the Aged competition designed to promote better relationships between elderly people and the younger generation.

Initially, old people were invited into the school for tea, cakes and a chat. Once they had got to know the pupils and felt relaxed, informal interviews developed. These were eventually videotaped using money awarded by Help the Aged and equipment from the school's central resource department.

Interviewing the old people involved a great amount of research beforehand. It's important to know in advance a bit about the background of the elderly people concerned. It was useful for the children to have a few questions written down to guide their interviews.

In such a way the project developed into a significant alternative curriculum item. Not only were the children gaining experience in videotaping and preparing interviews, but they were producing a valuable resource for use in the school and the community. At its height the young people spent two afternoons each week working on the project. As the Department had a blank/block timetable it meant that the pupils' lessons could be moved around to suit the project, although when the project was introduced into the school's regular History classes, restrictions were placed on the amount of time. This meant bringing the elderly people in at specific times and transport proved a big problem.

Clearly the project needed a co-operative attitude between the different Departments for it to work. Given this co-operation, the project was of enormous benefit to pupils studying "O" Level, "A" Level and CSE courses; they could fill out their projects with facts from the interviews.

Extra impetus was given to the pupils' work by the publicity the project received. Local television and radio companies, as well as the local press, expressed interest in it. The project appeared on the BBC and TV-AM. HTV's *Good Neighbour Show* allocated thirteen minutes to it. By reviewing the video tapes it was possible for the staff to write small articles for publication in the local press. These articles served the dual purpose of supporting the children's work by giving visible evidence of the importance of the project. They also jogged the memories of readers who might have had something to contribute.

The resources were not only used by pupils for examination purposes; much of the material was given to the local Heritage Museum. The School Resources Department also published a book entitled *Forest Memories*. Mr W L Freeman, an 87 year old who took part in the project, writes:

> The students were all so natural and interested. I was quite surprised at some of the questions and the answers they gave. They were certainly polite and well mannered, yet perfectly normal; although you would hardly think it would have been seeing the difference in our ages. And to my surprise I have found that even in the street they recognize and speak to me.

Giro

Another success story using video was the production of the video, *Giro – Is This The Modern World?*, shown on Channel 4 in July 1985. The video was produced in Telford through the Birmingham Film and Video Workshop by a group of young people who began working together when in the fifth form at a local comprehensive. This was their second programme; the first was broadcast in September 1984. It was called *What They Telling Us It's Illegal For?* It was based on the group's idea that home taping off the radio or records should not be illegal.

Filming an interview in the school canteen

Jonnie Turpie and Graham Peet were the adults involved. They got to know the Fifth form pupils when they asked if they would like to contribute to, and criticize, some poster designs for young people's rights. After meeting in school for an hour a week, the group began meeting every Tuesday night in the local community printshop. It took some weeks for the group to stop regarding one another in terms of a teacher:pupil relationship, but the intervention of the holidays helped.

The group also came face to face with the problem of money. *Giro* was actually produced to a very high standard through the working of the Workshop Declaration. This is a Trade Union agreement which allows ACTT (Association of Cinematograph, Television and Allied Technicians) members and non-professionals to work together on programmes available for broadcast. But these require budgets higher than those currently available in schools. Teachers not only face a problem in terms of resources, but also in terms of the time and informal atmosphere needed to produce videos like *Giro*. As Jonnie Turpie puts it:

> In the classroom it's difficult to criticise teachers, school, YTS or the government, or talk sincerely of personal concerns. Unfortunately, it's these areas, at least initially, that provide the bones of contention that need discussion if a basis for collaborative work is to be achieved.

Teachers are also under pressure to fulfil curriculum requirements. They have to hold together classes of 15 to 30 young people while restricted by a lack of staff and resources. This could be opened up by utilizing the staff, equipment and skills of other parts of the education system such as TVEIs, Colleges of Further Education, polytechnics, universities, local media groups, workshops and media centres.

Primary school work

Finally, the work of Vincent McGrath in London Primary Schools also helps to illustrate the wider applications of video in a school situation. Pointing out that the education authority must have between £1.5 to 2m worth of equipment in London's schools and colleges, Vincent McGrath has attempted to take the creative use of video in primary schools beyond the stage of documenting existing activities. Behind his work is the hope that, by using video imaginatively, children can come to a greater understanding of how programmes are produced, and why.

One of the video projects developed was a video news programme called *Assembly News*. Initially the programme was pre-recorded. It focused on national and international news. But the first programme was stilted and boring with the result that everyone decided to turn the programme into a live one. News issues became anything that was of interest to the children or affected their daily lives. The programme went out four days a week at 9.30 a.m. for a whole term and involved groups of children on a rota basis. Topics covered included consumer items on crisps, sweets, comics and cola drinks; news items included netball, football, swimming etc.; whilst interviews ranged from the school cook, the school secretary, parents and even a teacher who had just married.

Young children at the filming stage of a project

It was felt that the pupils made tremendous gains in a whole spectrum of inter-related areas, especially language and presentation skills. The project caught the imagination of the children. Many of those who had previously been uninterested in the more traditional parts of the curriculum suddenly became highly motivated. Happily, this spilt over into other areas of their work. The project not only taught the children to respond favourably to the pressure of getting the programme out, but also made them think about what questions to ask in interviews in order to obtain key information.

9. Video in the family

Home video equipment offers a family the chance to work together and to create excitement and entertainment. Similarly, for people who normally use video just at work, family groups – in or out of special occasions – can present the time and place for enjoyable and rewarding projects.

Most people experience video in the family through taping broadcast programmes off air or through the recording of family events. There is a small but growing number of video hobbyists who use the equipment on a rather more individualistic basis to produce home movies.

Home video equipment offers the family a chance to work together and to create excitement and entertainment

It's worth keeping to these three familiar types of home video project, and extending each of them into less obvious but more interesting areas of activity.

Recording options

The nation's personal video archive must consist of endless hours of weddings, barmitzvahs, birthday parties and holidays. Will our grandchildren and their children find these as fascinating as those sepia photographs handed down from our predecessors?

Whether or not the lack of editing or sensible restraint allows quantity to triumph over quality, if you use your video equipment simply for this sort of activity, you are missing all sorts of possibilities. The classic "family documentary" takes the family out of any sense of social or local context, ignores the riches and talents of individual participants and often catches people at their least creative and interesting. By concentrating on the excesses of the special event, you miss the richness of the ordinary and every day.

To make video as integral a part of the family archive as all of those photographs in albums and shoe boxes, you should also be recording other parts of the family cycle. Perhaps these appear less glamorous but in future years they will be just as interesting: redecorating or moving house; mending the car; a day in the life of the kitchen, and so on. Even great accidents and disasters can look wonderfully funny in restrospect.

A second possibility is to use video to concentrate on the individual histories of family members – both older and younger. This uses video as an oral and visual history record.

A third option worth exploring is to see your family in its context; where it lives (both its immediate neighbourhood and a wider local context); what work, or lack of it, family members do; hobbies; school; local history. These factors affecting our families'lives will provide a much more interesting set of records and memories than the formal set piece occasions.

Home editing and critics

The video recorder is the ultimate in convenience entertainment. Hollywood, Bombay and Elstree work 365 days a year to produce the big movies. The broadcast channels and satellites pump out programmes twenty four hours a day, seven days a week. We can watch any of this, anytime we choose by programming our VCRs.

One constant source of argument between parents and children is over the suitability of programmes. Rather than imposing parental standards in an unquestioning way, these skirmishes can be made to teach both children and adults such things as self-discipline, organizing a limited amount of time and deciding on priorities to name but a few.

One constant source of argument between parents and children is over the suitability of programmes

Yet the VCR is not just for copiers. It can also be used by editors and critics. Editors might take as a model any one of a number of compilation programmes (*The Best 50 Years Of BBC Sports, It'll Be Alright On The Night*) and put together a themed fifteen, thirty or sixty minutes for future viewing.

Critics, on the other hand, could take a day, a weekend or a week's broadcast output. Choosing manageable but short time-frames (5 to 15 minutes), plan a viewing schedule (start, perhaps, with one or two programmes only), and record sections of these which will show why you chose these, and what you liked or disliked about them.

The simplest way to organize this is to randomly record off-air programmes of one or two minutes length, leaving a similar length of time as unrecorded breaks between them. Then view the sections again, planning your comments by way of introduction and analysis, before going on to record them. The finished result might look like this:

- Caption;
- Introduction *(30 seconds);*
- First recorded section *(1 minute);*
- Analysis of first section *(30 seconds);*
- Introduction of second section *(30 seconds);*
- Second recorded section *(1 minute);*
- Analysis of second section *(1 minute);*
- Conclusion *(30 seconds).*

Home productions

Home productions made by and for the individual form the subject matter of a number of books and magazines. It is often more interesting to make a low-tech, high participation project using all the variety and richness of opportunity which a gathering of different generations offer at home or on holiday.

Whilst you may be able to do such a project with a relatively homogeneous group of children and adults, you are equally likely to find yourself in the midst of a group of people whose ages and interest spans the entire range. It makes the event much more exciting when everyone who wants to participate can be given, or found, a role both on and off screen. This goes for the under 3s as much as the over 80s.

Make a list of the jobs to be done – from holding the clapper-board or credits list to looking after the babies and animals – and make sure that everyone involved has something to do. Equally, be careful that people who don't want to take part aren't bothered by this media circus.

Be clear about who you want to involve in the technical side of the production. Everything that happens in school and youth club groups between older and younger, boys and girls, and assertive and nonassertive people happens as strongly (if not more so) within the family.

Undertaking such a project at home allows you to use everything and anything that you can lay your hands on from the kitchen, bathroom, bedrooms and living rooms. This means that you can organize and "dress" your sets and casts in a detailed and realistic way if that is the effect you are looking for. You will probably also have easy access to sound effects and music for sound tracks, make-up, costumes, etc.

Setting up a project

It is important to be quite clear why you want to use video at home and what you want to get out of it. It is equally important to understand what everyone else involved wants to do. You need to ensure that you neither impose your own motivations on others, nor proceed on the unquestioned basis that everyone shares your reasons and expectations.

In any event, of the stages to organize (however informally you manage this) the obvious ones are: preparation; rehearsal and practice; run through; recording; playback.

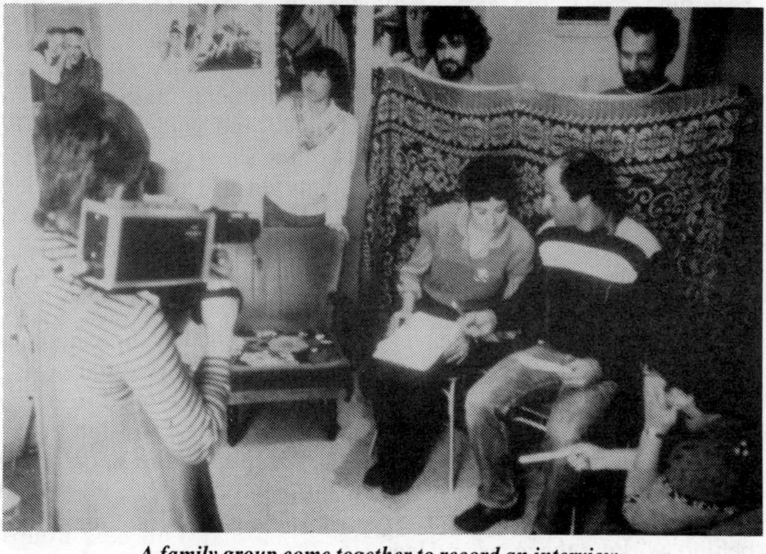

A family group come together to record an interview

You might want to start with one or two mini-projects and games to settle the group you are working with, to familiarize the participants with the equipment and to get everyone over the initial excitement or fear of seeing themselves on television.

Setting up a simple scavenger hunt may be one way of starting. This will enable you to sort out the practicalities of working around the house, garden or neighbourhood, particularly thinking about other uses and people before you start to shoot.

Video circus

One model project often used in family video is the circus. Structurally, this gives you a convention which is recognizable by children from an early age, a number of different component acts; the occasion for using costume, gags, tricks and illusion; music and colour; and a link-person (the ring master or mistress).

Because there are a number of different acts involved, everyone can be included doing an act at their own level or with their particular group of cousins or friends. This also gives you the possibility of involving a number of different combinations of people as the recording crews and also of allowing people to take part, both backstage and on stage.

The circus is a convention in its own right, and also suggests any number of parts: clowns, trapeze artists, jugglers, illusionists of all kinds, musicians and a variety of trained and wild animals. The running order, length and complexity of the show can be adapted to suit the needs of the group. Older children will often devise acts requiring fiendishly complicated special effects – part of the fun is working out how these can be done. As with most video projects, much of the excitement happens before and after the actual recording.

One summer holiday circus devised and recorded by three families (six adults, six children aged 8 to 13) included a nine year old sawing her brother in half; a twelve year old, whose head moved from the floor to the top of a 10 foot high pair of double curtains and back again; and a father and son double act. Dressed in the same baggy raincoat, they presented a dwarf performing on a table, the son in front with his arms dressed in a pair of wellington boots, the father behind working the arms and legs.

Everyone can be included doing an act

What not's

There are a number of things definitely worth avoiding when you're using video at home. Top of the list of what not to do is doing the right project with your family at the wrong time. Avoid birthday parties and feasts, as video tends not to go as well with jelly as cream or ice cream. And vice versa. Besides which, the attention span of most young people at events such as this will be much shorter – and the excitement level much higher – than you will need to have much fun.

This raises another question to do with the work, or perhaps group work ethic, and its place at home. The production cycle, and the equipment itself requires a certain amount of commitment and self-discipline. You will need to set up a clear set of rules to ensure fair play and the equitable, safe use of the equipment. You will need to dovetail this into the rest of the family's style and timing to avoid either casting yourself as the heavy handed uncle or aunt, or undermining the enjoyment of the whole affair by providing an excuse for uninvolved adults to undo the work by becoming unhelpfully involved, or removing their children prematurely.

When you have safely steered your way through this, and avoided the insertion of too many smarties or cheese straws into the machinery, you should have a cassette which the audience, participants and the rest of the party alike, will want to sit and look at. Provided, that is, you don't try to take over the television in the middle of the family's favourite programme!

10. Working with girls

All the other material in this book has been written as an aid to working with young people regardless of their sex. This section is about using video with girls on their own. Why?

There has been a growing recognition over the last few years that youth provision has been dominated by the interests of boys. Most club activities, such as billiards and table football, are traditional male preserves. The more aggressive behaviour of boys tends to exclude girls from taking an active role. These problems are rarely confronted, partly because staffing in the Youth Service is itself male dominated.

In the face of this, a number of girls' groups have been set up mainly with women youth workers. In this way they can form friendships with each other, gain confidence in themselves as independent people, and develop skills without the boys always taking on the leading roles that push the girls to the sidelines.

A girls' group on location

Deciding to use video

As always, it is important to be clear about why you're using video, when to use it and how it fits in with what the girls themselves want to be doing. One video worker meets regularly with a girls' group at a centre in London's East End. She is aware that, as well as being a place to work with video, the workshop has other important functions. She sees it as vital to recognize when they want to use video and when they want to spend the time being together and talking. It's somewhere that girls can talk freely to each other. Rather than cutting them off from a social life, the workshop is a place where they can meet boyfriends after having spent a constructive session.

The equipment

One of the advantages of using video with girls' groups is that they get to use the equipment. In a mixed group, the boys invariably take control of the equipment. It's not simply the boys' fault; often the girls themselves will defer to them because they lack confidence in their own technical abilities. The film and television industry is still relatively male dominated, and "cameraman" remains the accepted term – with union agreement – for camera operators of both sexes.

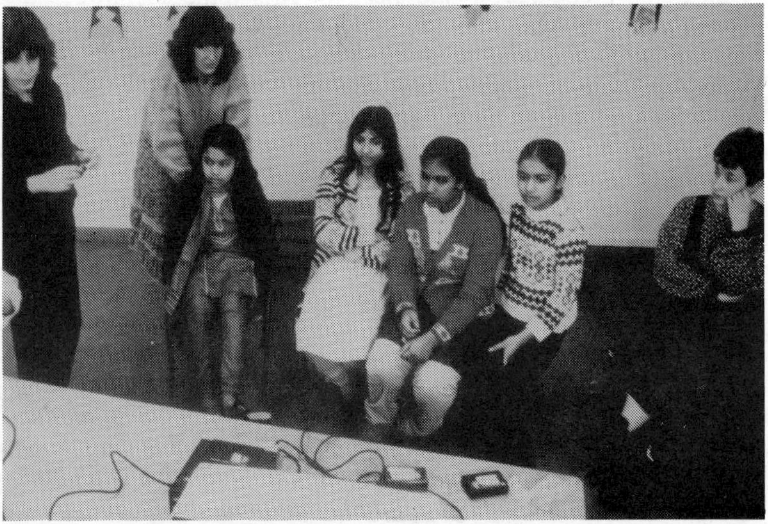

Getting to grips with the equipment

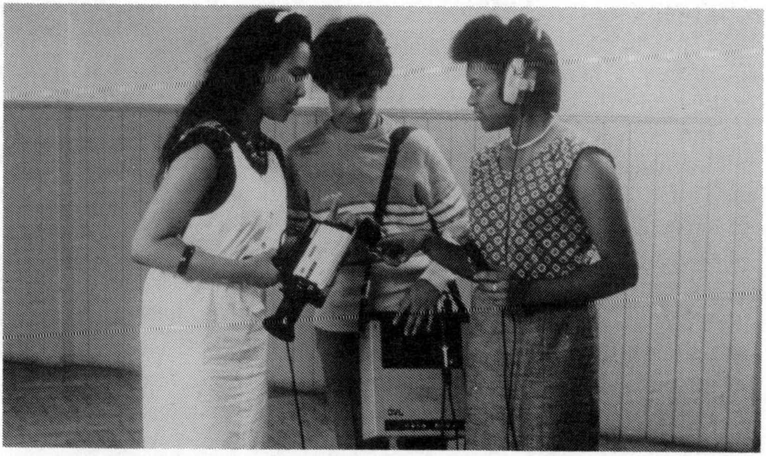

Operating the camera — no better way to learn

It's a common experience to hear girls who have used video say how they get a big buzz from being able to set up and operate video equipment completely on their own, because they did not believe they could do it.

For these reasons it's worth taking special care over teaching the use of the equipment in girls' groups. It can open up a general discussion about the way they view machines (reinforcing the myth of it being somehow difficult or beyond them) and how this limits their ideas on the sort of work they can do when they leave school. One girl in a youth club became interested in working with video after she had seen a female film-maker talking on television – until then she had thought it was something a woman could not do.

Seeing themselves

Girls tend to experience seeing themselves on video for the first time differently from boys. While boys exuberantly try to fill the screen with their own faces, girls will often turn their backs on the camera or hide their faces behind their hands. They're happier to appear for the first time in a group rather than on their own. As young women, they're used to seeing glamourized images of women in adverts, magazines and on TV. This conditions them to be concerned about how they look – almost to a level approaching anxiety.

It's important to be sensitive to this and to allow them whatever time they need to relax and gain confidence in seeing themselves on the screen. You could also discuss how the artificial images of women in the media make it hard for them to feel good about how they look.

One worker found a particularly effective way of working through this with a small group of very shy adolescent girls. She combined their first use of video with a hair-dressing session, a video camera and monitor making the ultimate mirror. As they were having their hair done they began experimenting with being on screen, putting on smart voices and pretending to be film stars for the camera. By the end of the afternoon they looked really good, and had lost many of their inhibitions about being seen on the monitor.

Projects

When you get around to planning a project you may notice more differences in approach between boys and girls. If they're working on a drama or role-play, girls tend to choose more personal themes, like relationships, pregnancy, family or school life. While boys might go for TV-inspired fantasy, girls seem more interested in exploring topics closer to the reality of their own lives. It's useful to bear this in mind if you see part of the purpose of the group as providing support for girls questioning their social conditioning. Of course, this is only a generalization and has many exceptions. One girls' group on a playscheme worked out and recorded a fast-moving murder story, complete with aggressive *Sweeney*-type cops and a big fight at the end.

Many of our ideas and feelings about the place of women in society, in relation to men or the family, are reinforced by the way that these themes are treated in the mass media. Video provides a tool with which to confront these concepts, although other resources (newspapers, magazines or photos) can also be useful. One youth worker spent an evening with a girls' group cutting up and re-arranging adverts from Sunday colour supplements. They made some powerful discoveries about sex roles by such simple actions as putting men's heads on women's bodies (and vice versa). They then discussed the resulting images.

You could do a similar sort of project on video, for example, by recording an episode of *Eastenders* off-air, deriving a script from it and then re-enacting it on video with girls taking all the roles, both

Action! A girls' group records a playground scene

male and female. Or you could videotape an advert, such as a soap or food commercial, with a woman in very traditional role, and then work out and substitute your own dialogue. What do the group think her feelings might be if it were not an advert?

One girls' group used themselves to explore how television goes about glamourizing women. They first recorded their whole bodies, then their faces and finally did just a close-up of their eyes. By using different music and soundtracks, they learned how the mood or perception of the images could be altered. Finally they added a spoken sound track to end up with an advertisement for cosmetics.

Girls' groups and boys

If you're working in a club or centre that is open to both sexes, the chances are that the boys will feel threatened by a girls' group. They might try to disrupt it by invading the room you're working in, and they're likely to be contemptuous of the video work the girls have done, putting it down to play on the girls' lack of confidence.

One worker used to play back tapes made by both girls and boys to all the members of the club where she was working. She found that the boys never bothered to look at what the girls had done (although everyone had watched the boys' tape); and that the girls themselves, without any good reason, assumed their own work to be worse than the boys.

So it's important to find a secure place for a girls' group to work in and to select the right time to show their work outside the group. The unconscious selectors in education and the Youth Service still very much favours boys although, increasingly, workers are trying to redress this imbalance. To find out more about working with girls see the end of Appendix 3.

11. Playback

Showing tapes

Most of us watch TV in a very passive and uninvolved way. It washes over us, out of our control – apart from switching off, or changing channels. Consequently, it's hardly surprising that young people's enthusiasm for video comes first of all from being able to show something on the screen that they have made for themselves.

You can break away from the conventional patterns of viewing at home, both in the sort of material you choose to show and in the way that you show it. It is increasingly possible to obtain video tapes made by community or youth groups about areas of special interest to, or educational use with, young people. These tend to have been made by people from their own experience and often speak more directly to viewers than conventional television programmes do. This is especially true if the audience for a tape is from a similar background or share experiences with the makers.

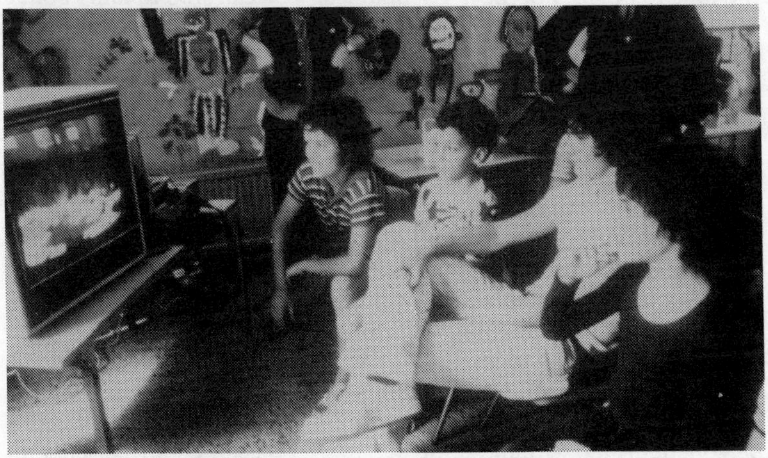

Viewing the take – what you shoot is what you see

There are increasing numbers of tapes available that have been made by young people. Some concentrate on issues facing young people today. *Giro – Is This The Modern World?* was made by a group of young people in Telford in the Midlands who described it as:

> ...an investigation into work and the work ethic, doing something you enjoy and getting paid for it; the level of the dole and why young working class people have only the choice of £22.45 a week on the dole, £25 on YTS or £40 for working in a factory.

Two young women, who were part of the team that made the video, write:

> Young people all over are taking an interest in their environment, culture and politics and are wide awake and letting others know about it...*Giro* is very much a young person's point of view – come to Telford and ask some of us.

Giro's reviewers seemed to agree. *Time Out* wrote that "it reaches a lot of those parts that more formal, 'professional' documentaries don't even come close to."

Even if you're playing back a recording of an ordinary broadcast television programme there are ways of getting the young people more actively involved. It's the watchers who control when you want the tape to begin, not the TV schedules. You can stop to talk about a sequence the group is particularly interested in, skip those which they find boring, rewind and see parts, or the whole, again.

You can make the viewing experience more active in other ways too. Suggest that the young people act as TV critics for a newspaper and write a piece on the programme they have just seen. Alternatively, design a poster for it. They could add another soundtrack over the top of the original. If you can get hold of editing equipment, they could plan and make a trailer for it by selecting shots from the original and adding their own soundtrack. They could even re-make a sequence from a different point of view.

If the object is to stimulate discussion of the subject matter of the tape, it's a good idea to invite someone along to the showing who was involved in making the tape or knows about the issues it deals with. When the tape is finished, it's rather easier to have a discussion with a person identified with the subject matter. You can't hold a discussion with a blank TV screen.

It's a good idea to invite someone along to the showing who was involved in making the tape

One worker arranged for this to happen with a group of 14 to 16 year old boys, about half of whom were regularly getting into trouble with the police. The worker felt that the young people could probably use more information on the prison system on the grounds that they were certainly putting themselves in danger of ending up within it.

They obtained some video tapes from PROP (Preservation of the Rights of Prisoners) about a tribunal on the trouble in Hull Prison in 1976. They also learned that one of the prisoners who had testified at the tribunal and appeared on the tapes now lived nearby. He was invited to the group on the evening of the playback. The tape ran for about forty five minutes and was visually very boring; just one continuous shot of him standing talking into a microphone, but the group watched fascinated throughout and stayed talking to him for over an hour afterwards.

Part of the reason for the success of the evening was the nature of the information that the young people gained from the video tape – a direct and raw account of the prison system. But it was also very important that the ex-prisoner, having appeared on the tape, was there in person with the group. It made the connection between what was happening on the screen and "real life" more vivid than could ever be realized watching the programme on television at home.

A similar sort of thing happened when a female youth worker arranged for a women's aid organization to show a group of girls a tape they had made. After they had seen the tape, a few of the women from the refuge talked with the girls. Some of the girls were able to talk for the first time in the group about ways in which their fathers acted within their own families. These were experiences not very far removed from those that the women on the tape had gone through.

Obtaining tapes

Getting hold of cassettes does require some initiative. Most schools, Adult Education Institutes, Colleges of Further Education and resource centres now have video as part of a range of audio visual aids. They record programmes off-air for use within their institutions. If you cultivate the right sort of relationship with a technician in one of these organizations you may be able to borrow their material or even have them record copies for you.

Also, a number of youth offices and education authorities now have libraries of suitable material for use in their catchment areas. An increasing number of distributors are serving the youth and educational markets and it would be worth contacting them and obtaining their catalogues. See Appendix 4 for a list of organizations. Obviously, the recording and replaying of programmes for a charge or selling privately recorded tapes is piracy and is illegal.

12. Editing

In the early stages of using video with young people, it's usually a good idea to try to make a completed tape within one or two sessions. In this way the group quickly gains a sense of achievement. But to make a tape which looks complete and interesting always takes a certain amount of editing.

The simplest form of editing is called editing in camera. This means that you're selective in what you shoot and you shoot it in sequence. In other words, you need to plan out the shots you need and then shoot them in the order in which you want them to appear in the final programme. If you want to use titles, then these have to be prepared and shot when you want them to appear. Any shots that go wrong have to be recorded again before going on to the next shot. You can record the soundtrack as you go along, or decide from the beginning to dub all the sound on afterwards.

Fine editing on a U-matic editing suite

There will be times when you can't work like this and have to shoot out of sequence. A documentary project, for example, will often involve interviews with people outside the group which usually can't be organized tightly enough to shoot in sequence. This means that you will have to edit the material you have shot, both in terms of eliminating the bits you don't want and putting it all into the right sequence so that it makes an understandable programme. Also, if you're working with a group over a longer period, they may get bored with simple recording and want to learn how to edit.

This section deals first with how to edit and secondly with how to involve young people in the process.

Young people might be bored by repeated viewing of the same unedited material

How to do it

It will save a great deal of time and effort if, before you sit down with the editing machinery, you have played back and studied in detail the tapes you'll be using. It helps to have logged the material into a list of shots with brief descriptions of the content of each one. In this way you'll be able to find the bits you want quickly. Work out which shots you want to go where, which shots will look and sound right together and any inspired ideas you may have.

This is not the place to go into technical detail about editing, nor about the different systems and types of machinery around. All of them follow the same basic principle of transferring material from one VCR to another. The two VCRs are linked by leads taking audio and video signals out of the first and into the second. You play back the shots you want from your original tape on the first machine and record them in the sequence you want them to appear onto a new tape on the second machine.

The best way to learn how to edit is to find a place that has the right equipment and someone who will give you basic instruction. There is usually a community video workshop in most towns and cities and there are courses at most Evening Institutes. It's possible to get unreliable, but acceptable, results with two ordinary domestic VCRs – a technique called crash editing. The skills of editing are very easily picked up with a little practice.

Editing with young people

From what has been said so far it should be clear that editing is not a suitable or attractive activity for everyone. The planning stage requires a lot of concentration and commitment to overcome the boredom of repeatedly playing the same bit of tape to find the exact shot that is wanted. This has none of the immediate appeal and excitement of shooting; it is certainly not a natural group activity. The physical task of operating the editing machinery can really involve only one or two people at a time, so the rest of the group has nothing to do.

It would be wrong, however, to give the impression that it's impossible to involve young people in the editing process. Some find it a very absorbing and creative activity indeed. Two teenagers, who were particularly committed to a tape they were making, quickly learned how to operate the machinery and spent a whole day completely by themselves editing it.

There is a way in which a larger group can become involved. Perhaps assisted by one or two interested members of the group, you can go off and do a rough edit to show the whole group what the final tape might look like. Though young people might be bored by repeated viewing of the same unedited material, often they will look keenly and critically at a rough edit of material they helped produce. They are usually very sensitive to the difference between a good cut and an awkward one, perhaps through watching TV. After showing them the rough cut and discussing it, you can then go ahead to make the final edit according to the instructions of the whole group.

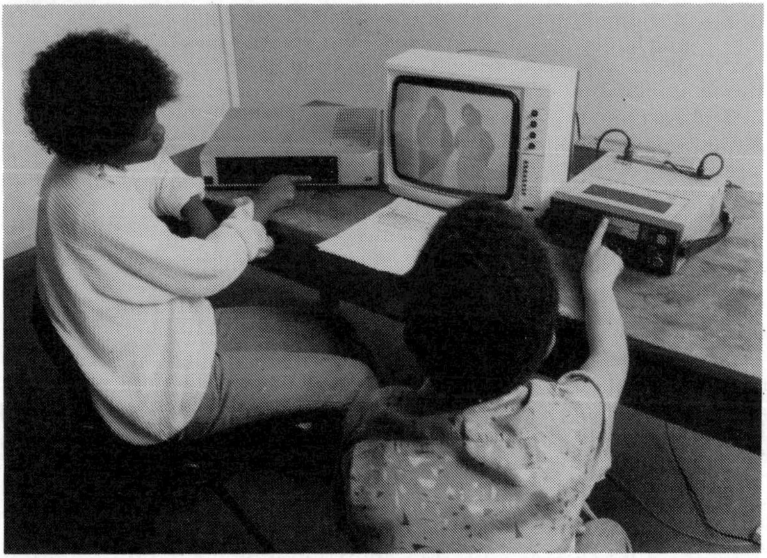

Two young people line up shots for editing

13. Video as part of live events

This section describes projects where video has been used beyond simply recording and playing back. They are offered as examples to modify for your own use. You may find them particularly useful, for example, as frameworks for one-day projects on playgrounds and playschemes or as part of community festivals.

The man from Monochromous

This one day event was first staged by a group in South Wales at local playschemes. In it, video acts as a stimulus for imaginative craft work and drama activities, as well as giving credibility to the idea of contact with an alien being. The day begins with people setting up video equipment in the hall where the playscheme is being held. They say that they have come to establish a final communications link with another planet. The hall, apparently, is the only place where good communication can be ensured.

The young people are puzzled and gradually gather around the monitor. Suddenly a strange alien being appears on the screen and they begin to ask it questions. It slowly emerges that this creature lives on the planet Monochromous and is planning to visit Earth. The creature is worried about how to become acclimatized to the new environment, particularly because there is no such thing as colour on Monochromous (the colour on the TV monitor is turned down for this).

The young people undertake to build and paint environments which will introduce the alien to different colours and other aspects of life on Earth. The alien seems satisfied and says that he should reach Earth "by your after-the-noon".

The picture on the monitor fades and the young people begin making preparations for his arrival. They build a safe area out of newspaper to quarantine him before he can be released into the world of colours. After lunch they make contact again. The alien is aboard his spaceship and will soon be arriving on Earth. Preparations

Preparing to welcome the man from Monochromous

continue at a more hectic pace. Then contact is made yet again and the countdown to his arrival begins with the young people joining in.

On "Zero!", the alien beams himself down through the serving hatch in the corner of the hall. Some groups put on performances to celebrate his arrival and then escort him around the various environments, acclimatizing him gradually to life on Earth.

It was important that the young people and the actor playing the alien could see and hear one another. Technically this was achieved by using two VCRs and cameras in closed circuit. They were connected over a distance with two monitors.

Camera and VCR "A" relayed the image and voice of the actor (hidden away in a separate room) to the monitor where the young people were gathered in the hall. They, in turn, were picked up on VCR "B" and relayed to a monitor in the actor's room.

Space Monster

This was a project devised and staged by a group of London video workers to entertain young people at an Open Day held at their workshop.

They used the same technical device as "The man from Monochromous" event: two video systems in closed circuit, connected to distant monitors. They built a space monster in the yard outside their workshop. It had a spherical-shaped television as its head. There were two tentacles with a microphone at the end of one and a camera on the other. The television "head" displayed pictures from another camera which was pointed at an actor back in the workshop studio. On his monitor he could see and hear what was going on in the yard.

The way that the young people were involved was less formally structured than in "The man from Monochromous", though based on the same idea of exploiting the differences between Earth and life on the space monster's planet. The monster, for instance, didn't understand what clothes were or the distinction between the sexes. Later on they had a dog in the studio which appeared on the monster's television "head" – so some young people went to find a cat to show to it. Towards the end of the afternoon the studio was discovered by other young people who took over the whole set-up for themselves.

TV Man

TV Man was a pre-recorded tape made by a community arts group to be used interactively with young people as an introduction to working in video.

The tape begins with a picture of a test card which suddenly falls away to reveal a man in a small room cluttered with equipment and gadgets. He introduces himself – he has been trapped behind the test card for seventeen years by "them" to make sure the TV programmes go out.

He asks the young people to do some stretching exercises and then to breathe on the screen which then goes blurred. They're then asked to help wipe it clean with their sleeves. He realizes that he could be free if the young people could prove that they didn't need him to produce TV programmes, so he asks them to copy everything he does.

He makes a series of weird noises and complicated movements, which they follow, and then asks them to copy the person furthest from the screen. After a while that person has to touch someone else and everyone then has to copy the second person. They do exercises to help them if the sound goes wrong; they have silent conversations in pairs by just mouthing words. Then the pairs split up and begin to speak to each other as loudly as they can from opposite ends of the room. Next, they have to mime pictures to the sound he makes such as horses' hoofs or the noise of someone sawing wood. Then he initiates a story ("There once was a man who escaped after seventeen years in a box...") which has to be continued by someone he nominates ("You there, first row, third from the end. Carry it on.") who passes the story-telling role on to the next person. When the rest of the group have added their bits to the story he sets them a final test – to make their own fifteen minute TV programme.

You could make a tape of your own along similar lines using the same idea of getting young people involved in activities upon the suggestions of a character on the screen. Once you have a tape like this you can occupy a large group with only a VCR and monitor.

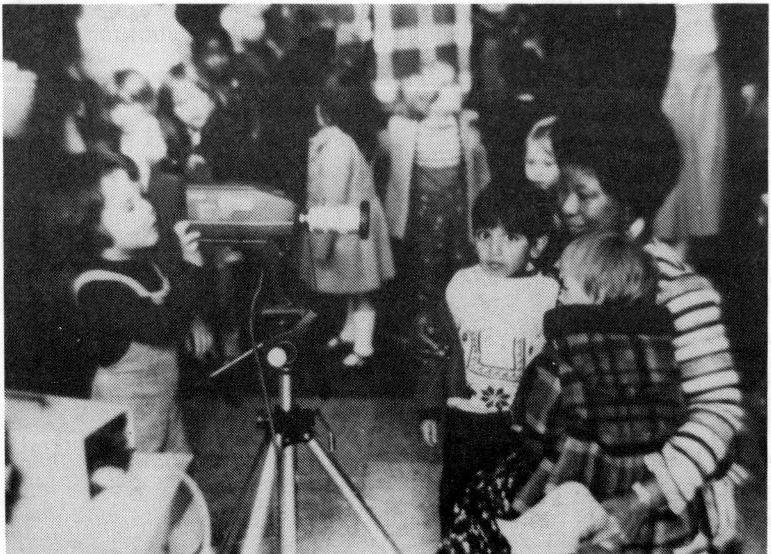

Activity for everyone — making mini-programmes

14. Distribution

This section is about finding ways of showing the tapes you've made with young people to a wider audience.

Which audience?

In a lot of video work its value often lies in the involvement and satisfaction a group gets from using video and playing back the results to themselves. It doesn't need to be justified by showing it to outsiders.

However, there are times when it's important to show tapes. In some of the work described in the chapter, "Projects in The Community", for instance, the purpose of the projects was that the work would be shown in the community and to adult authorities.

It can be important to show work in other circumstances too. If a group is going to be using video over more than a couple of sessions they may become bored with having only themselves as an audience. Knowing beforehand that work is to be seen by outsiders can motivate them to find out more about the medium and what can be done with it. It can also provide a reason for continuing after the initial excitement of seeing themselves on TV has worn off. Also the feedback received at a showing from people who weren't involved in making it gives a better sense of what they have achieved.

But be careful. Much video made by young people is only of real interest to the people who were involved in making it. Showing it in circumstances where it may get an uncaring or even hostile reception will obviously damage the confidence of the programme makers and, possibly, their will to learn more.

The best way to be sure whether or not to show a tape is by being sensitive to the feelings of those who made it. A group once spent about six months working on a tape with a science fiction theme. It turned out to be technically sophisticated and quite polished. The

Young people view their work before deciding which particular audience to show it to

video workers involved were keen to set up playbacks of it in the local community centre and perhaps also in the school which some of the group attended. But in the end the group vetoed these suggestions, feeling that it would be a terrible "show up" if any of their friends saw it. They had grown up a lot while the tape was being made and no longer felt comfortable with the way they had been presented in it. So, whilst they felt happy about it being shown to groups of interested adults, it quite rightly never had any public showings to other young people in the area where it was made.

Getting shown

Once your group has decided that a tape ought to be shown you then have the problem of how to do it. At a local level, you will have to find, or create, your own outlets suited to the kind of programmes you want to show. Here are a few possibilities:

- Organize a competition open to schools and youth groups in your area and have a day where all the entries are screened. This would depend either on there being a number of groups near you who are also using video, or upon your organizing it well in advance so that inexperienced groups who wanted to be involved could make tapes in time;

- Start showing tapes in other youth clubs and centres in your area, gradually building up a distribution circuit. Again, this would work better if yours were not the only group in the area using video. With more groups using it you could start a tape exchange network between clubs;

- If there is a community festival of some kind in your area, set up a stall or tent and show your tapes;

- Most schools now own video equipment. Some even have a cable linking several classrooms for showing tapes. You could try to persuade one nearby to show your tapes in breaktimes or, perhaps, as part of a social studies course if the tapes are relevant;

- You're very unlikely to get tapes shown on broadcast TV, because the broadcast standards are very high indeed and because of union complications. Even the BBC Community Access programmes, which have guidance by the group concerned, are made by BBC technicians. If you have made an exceptional tape you may be able to convince a news or teenagers' magazine programme to do an item on it. However, be very clear about why you're doing this and ensure you have a degree of control over how the tape is used. TV people always have the interests of their programmes at heart and these could conflict at some point with the interests of the young people you work with. In exceptional circumstances it's possible for a group outside television to make and control their programme.

Conclusions

Young people tend to be more interested in making tapes than in getting them widely shown. This is partly due to a lack of confidence. For these reasons it's important that they should feel good about where and to whom their tape is shown. If possible they should be there to operate the playback machinery and to see and hear the reaction of the audience.

With home video equipment now so widely used, it's not too difficult for young people to organize their own showings. It usually does wonders for their self-image when family and friends value what they have done.

Don't forget that the most important audience is most often the young people themselves. Whatever the apparent quality of the tapes, it's a good idea to retain them for as long as possible. A group of young people can get a real sense of their own history from looking at a tape they made a year or two ago.

It usually does wonders for their self-image when family and friends value what they've done

Part 4: Equipment

15. Borrowing equipment

Borrowing equipment is not as easy as it should be in an equipment rich society. There is a wealth of equipment lying around. More often than not it's scandalously under-used. Educational institutions, Local Government offices, particularly in Social Services, Planning and Recreation Departments and other such places often have hardware which might be available if you work there or know someone who does.

Borrowing institutionally owned equipment is obviously more difficult if you're not part of that institution. Personal contact is the only way one can realistically hope to gain access. If you don't know the right people, you have to get to know them. Involving a teacher or a technician from such an institution in your project can often work wonders.

With this type of approach you will need to be sensitive to people's protective instincts when equipment in their charge is

Personal contact is the only way one can realistically hope to gain access

involved. They may have insurance to worry about. The idea that young people are able to use video in a worthwhile and responsible manner might be new to them. Putting yourself in their shoes is a good place to start when looking for ways to allay their fears and to convince them the equipment would be safe and well-used.

Resource centres

A number of media resource centres have been established in communities all over the country. Some have been set up with an educational brief, others with the purpose of making the media available to groups who would not otherwise have access. Some of these centres loan equipment free; others make a small charge (usually well within the range of youth centres). Even if they do not own suitable equipment themselves they might be able to tell you where to get it and offer general advice and support.

The list in Appendix 3 is not exhaustive as centres are opening and closing all the time. The Regional Arts Association for your area will know of any local resources, and in some cases loans out equipment itself. What is available varies widely, but facilities fall into four basic categories:

- Portable shooting equipment;
- Editing suites;
- Training facilities;
- Specialist equipment (microphones, audio mixing, etc.).

16. Buying equipment

Buying video equipment for your project can be a daunting and difficult task. There are several different systems available, and several different varieties within these systems. You'll find yourself up against a wide range of models and a wider range of sales claims, as well as the rapid development of new models.

This section outlines the equipment you need, its likely range of cost and the advantages of the different models available.

There are usually quite a few dealers prepared to compete for your custom

Even if you're not buying and not particularly technical, this section is still of use to you. Familiarizing yourself with the equipment through browsing in shops and going to exhibitions can be very helpful in answering technical questions raised by young people.

Test shot. A young girl tries out the equipment

If you're the person buying equipment but are not technically confident, then make sure the person selling it doesn't sell you something you don't need! It's your money, so get clear answers to your questions. There are usually quite a few dealers prepared to compete for your custom and you can often get a good discount if you can pay promptly. Therefore, always get at least two opinions and two price quotes before you make a move. Even better, of course, would be to ask someone who is running an educational resource facility for advice.

The systems

At present, there are four main systems that you need to know about: U-Matic, VHS, Betamax and the new 8mm format called Video 8. They all use cassettes of different sizes so you cannot play back a cassette recorded on one system on another type of VCR. As a result, your choice may be dictated by the type of equipment your local authority or community centre uses so that you have compatible equipment.

As a general guide VHS now has at least 70% of the domestic market, Betamax 25% and other systems 5%. There are currently seven million video cassette-recorders in the U.K. – one for every three households. In the field of community work, VHS and Betamax have a more equal share of the market. However, it's possible to transfer material from one system to another to enable the widest showing of your tapes. There is a loss of quality, however, of approximately 5% to 10% each time you do such a transfer.

VHS Like Betamax, VHS uses a cassette containing half inch wide tape. It was originally designed to compete with Betamax in the home market. VHS stands for Video Home System. So, at first, it was mainly used to record TV programmes and to playback tapes from video shops. It has since been strengthened and modified for use outside the home. Additionally, there is now a VHS-Compact. This is a lightweight portable cassette system which runs for thirty minutes and contains a small VHS cassette. It is considerably less bulky than some other portable systems. A special adapter enables you to use the cassette in a normal VHS VCR.

Betamax This is the other half inch cassette system. It produces better quality picture than VHS but costs more. It's very much less popular in the U.K. domestic market than VHS. If somebody brings in their own VHS cassette you won't be able to play it on the Betamax.

Video 8 The new 8mm system began to appear in the mid-1980s. It uses tape only a third of an inch wide in a cassette just slightly larger than an ordinary audio cassette. This 8mm system has apparently been accepted by all the major manufacturers and looks likely to become the standard in years to come. But it first has to replace an enormous number of existing machines and this may take some time.

U-Matic By far the oldest cassette system around, this uses three quarter inch wide tape in cassettes ranging from ten to sixty minutes in length. The system was originally designed for producing relatively high quality productions and can be used for very accurate editing and sound mixing. But be prepared for a cost about three times as high as VHS or Betamax. U-Matic now has two versions – ordinary "Low-Band"and the very much more expensive, broadcast quality, "High-Band". It's very unlikely that you will come into contact with "High-Band" unless you get involved in broadcasting the programmes you make.

There are a few other systems which have been produced in the last few years and are now rarely in use. VHS, Betamax and U-Matic are the three main systems you will come across – with Video 8 possibly coming in to join the big three. Future developments could see a decline of VHS and Beta, and the introduction of a new Videodisc which can be used to record on.

The costs

The figures below are all based on average prices in the mid-1980s at the lower end of Betamax and VHS equipment. For a complete portable video kit you will need:

	Guideline prices (mid-1980s) £
Camera	£ 500
Portable VCR	£ 500
Mains unit/battery charger	£ 50
Accessories:	
1 spare battery)	
1 light)	
1 microphone)	
1 tripod)	£ 200
Colour television	£ 150
Total	£1400

Instead of buying a mains unit you could get a tuner/adapter which fulfils the same function and also allows you to record programmes from broadcast television.

As new models are being introduced all the time, prices do fluctuate considerably. But this also means it's sometimes possible to buy very good, but discontinued models at substantial savings. You can also buy used equipment, but be very careful about the state of repair of what you're buying. It's a bit like the risks involved in buying a used car! Will the seller give you a guarantee?

Furthermore, Betamax, VHS and Video 8 all now produce "cam corders" (the camera and recorder are combined in one unit) for around £1000. This is a useful option to consider if you're not interested in recording TV programmes.

A final word of warning: it's important to buy from a reliable dealer who knows about video equipment and how to get it serviced (even if it means paying a little more). Shops which do not specialize in video will tend to have no immediate back-up and repair facility which may not be very helpful if your equipment breaks down.

Things to look for

The technical specifications on video equipment can be confusing – even intimidating – to someone who doesn't understand technical jargon. Sales people sometimes use the information to blind you with science rather than to enlighten. To avoid this happening to you, here is a simple guide to the technical bits:

Cameras You will come across several different types of camera. The main factor which controls the quality of picture is the tube inside each camera.

A low-cost high-quality video camera

Tubes How many, how big and what kind are they? Ordinary cameras have a single tube. For more money you get cameras with three tubes, one for each of the primary colours. Good camera tubes are two thirds of an inch in diameter; cheaper ones are smaller at half an inch.

The basic kind of tube is a "Vidicon". Twenty times more sensitive to light than Vidicon tubes are the "Newvicon" type. These are designed specifically for low-light capability but the colour and picture quality produced are poorer in more normal or well lit conditions. An even better all-round tube that can be used in a wide variety of lighting conditions is the "Saticon". Ignore the brand names and variations, because if the camera doesn't say Newvicon or Saticon it's very likely to be no more than a modified Vidicon.

A newer type of camera uses no tube at all, but an electronic "image sensor". This is usually referred to as a "CCD" or "MOS-chip" or by some trade name. It's probable that these will improve in quality and become more popular as the cost is reduced. They allow cameras to be smaller, lighter and more robust than cameras with tubes.

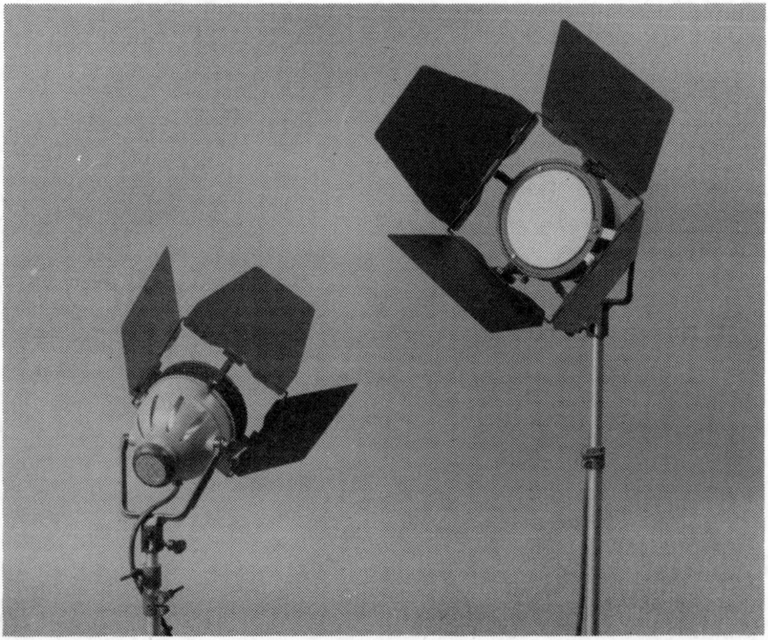

Video lights with barn doors

Light Required The amount of light needed for shooting is measured in "lux". Ten lux is a very, very low light, not much more than the light of a candle. Forty or fifty lux is a dim room. The level of lighting for a brightly lit room is approximately 900 lux. Manufacturer's figures for minimum lux, namely the minimum amount of light the camera needs to operate, are little more than a figurative indicator, so rely on a demonstration rather than on lux numbers.

The Lens Cameras are usually equipped with zoom lenses which can often be operated automatically. But be careful: automatic zoom and focus can often take away your control of the camera and you loose artistic and creative control of the lens. The amount of light that enters the camera is controlled by the iris or aperture on the lens which operates very much like the iris of the human eye. It becomes larger or smaller to let in the right amount of light. Most cameras have an automatic iris/aperture which adjusts instantly to the light conditions.

Some cameras are equipped with a manual override. This means that the camera can be set precisely in order to obtain the best image under conditions which would confuse the automatic system.

The aperture opening is measured in "f-stops" which can be unnecessarily daunting. The numbers make sense once you get used to the fact that the smallest number refers to the largest opening. The small numbers used, like f-1.4 or f-2.8, indicate the largest physical size of the aperture, thus letting in the most light. Large numbers such as f-11 and f-16 denote a tiny aperture which will only let in a correspondingly small amount of light. This setting is used in bright conditions.

The relationship between the numbers is that each stop as you go up from 1.4 halves the size of the aperture and, therefore, halves the amount of light it will let through. So f-4 is half the size of f-2.8. This can be demonstrated by looking into the lens and moving the iris control or, better still, looking into the lens of a 35mm stills camera and moving the iris control.

A camera with autofocus is very useful when working with young people as there is one less control to fiddle with. It's also a great help when shooting fast action, such as sport, which can be difficult for anyone to keep in focus.

Colour Temperature Although the human eye doesn't notice it, the predominant colour of light varies a great deal according to the source. Daylight is slightly blue in colour and ordinary bulb light is yellow. This is described by the term "colour temperature". Video cameras compensate with an appropriate filter. On some cameras this is a switch, on others a filter wheel. If you will be shooting regularly under fluorescent lights, which produce a sickly greenish glow on video, note that some cameras do feature an appropriate filter in addition to the daylight and ordinary light bulb settings. You can also buy a separate screw-in filter for fluorescent light which goes on the front of the lens.

White balance Some video cameras have to be told what the colour white looks like under different light conditions so that all the other colours can be automatically adjusted accordingly. This is called the "white balance". A camera with a white balance control has to be set each time it's turned on. White balance is sometimes set automatically once the colour temperature has been set.

Camera microphones It's worth asking questions about the type of microphone that is built into the camera. There are two kinds.

"Omni-directional" types pick up sound indiscriminately from all around. The other is the "cardioid" or "uni-directional" type. This picks up more of the sound in front of the camera than behind it. Again, your choice will relate to your working conditions. If the space has an echo or suffers from unwanted noise (like traffic, people in other rooms, etc.), then the uni-directional is the preferable type to use.

Having said that, it's best to have a separate microphone on a lead so that it can be placed close to the sound source. Built-in camera microphones pick up everything the camera person is saying!

A final item to consider with cameras is a bit of a luxury. Many cameras offer a built-in character generator. This lets you add titles or the date, place, names or other such information over a picture.

Asking all these questions about the camera you're thinking of buying is useful. But, before you jump, try it out before buying. See how it operates for yourself. Does the camera suit your needs and can you afford it?

Video cassette-recorders Portable VCRs especially those for VHS or Betamax, are often sold along with a particular camera. Some cameras and VCRs are not inter-changeable without different plugs and rewiring.

Portable VCRs have all the controls necessary for play, fast forward, rewind, recording and audio-dubbing on them. In addition, all compatible cameras and VCRs have a facility for making the VCR start recording from a button on the camera and many VCRs have a video search facility.

Some video recorders also have an output for independent video and audio signals. These are useful for editing. On the small format VHS VCR there are only outputs. On larger VHS and Betamax there are output and input sockets. This means the machine, when coupled with another VHS or Betamax, can be used to do rough or "crash" editing.

Most modern VCRs have a facility whereby each time you stop and then start recording again, the picture cuts in cleanly. Check that this is so on the machine you want to buy. If it is, then it will also give good clean edits when recorded on from another VCR.

The small format VHS portable has the advantage that it's very light to carry. Similarly, a Betamax machine fits into a container with the camera which then can be slung over your shoulder. Slightly more complicated and larger VHS portable recorders are heavier but not uncomfortably so. U-Matic portable machines are very heavy and can become tedious to haul around when your group is getting tired.

As with cameras, before you choose a VCR, have a go with it, carry it around, see how easy it is to operate and what special features are available.

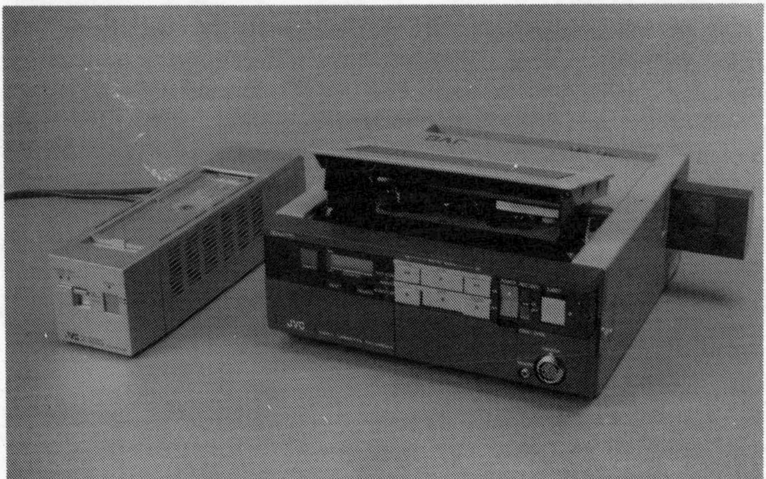

A portable VHS cassette-recorder with its battery and power supply

Televisions and monitors It's important to understand the difference between an ordinary TV set and a monitor. A TV set is designed to receive broadcast programmes or a signal from a VCR. This signal travels through a single cable from a television aerial or a VCR into a socket on the back of the TV set. Monitors, on the other hand, are only designed to run from VCRs and use a special video signal and video socket and an independent audio signal and socket. These give a higher quality in sound and picture. A good combination is the TV receiver/monitor which does both. It enables you to use the aerial socket, either to receive TV broadcasts, or to connect to the VCR and playback from a cassette. TV receiver/monitors have special audio and video inputs that are connected to the VCR.

Straightforward monitors, however, do have extra circuits so that you can loop through. This means you can have several monitors, one after the other, showing the same picture from a single VCR.

Some monitors and televisions also run off batteries so that they can be used anywhere. This is a considerable advantage for work with young people.

Televisions and monitors come in many sizes and weights – from six inches to thirty two inches measured diagonally across the screen. Probably one of the most common sizes is the fourteen inch portable television. Whilst it doesn't have the separate audio and video inputs, it is lightweight and can be easily tuned into your VCR.

Tripods When selecting a tripod, it's important to get one that fits your camera properly. There are three different types of heads on tripods – the head being the section that the camera sits on and which can be tilted up and down and moved from side to side. The highest quality tripod has a fluid head and moves very smoothly. It's unnecessary for all but the heaviest camera and you will never be likely to use it in community video work. The fluid-effect head is a more useful tripod. It can be bought fairly cheaply and gives a reasonably smooth movement. Finally, purely mechanical heads are cheap, but with a small lightweight camera on them the movement can be a bit jerky.

As with the other equipment, try out different tripods with the camera of your choice. The more stable and sturdy the tripod, the smoother will be the movement of the camera.

Lighting Using extra lighting in video work can make a lot of difference. Even a couple of ordinary light bulbs turned on can brighten up a scene. Modern cameras are very sensitive to low light conditions and work well, even with one strip-light! Then addition of a couple of bulbs with basic reflectors can change a scene enormously.

You can also buy lamps on stands fairly cheaply. Don't buy anything too expensive. Experiment first with different angles and effects. Working with lights can be very rewarding but they're not absolutely necessary.

Batteries You cannot go very far from mains power without batteries. Spare batteries are virtually obligatory with young people – a battery has a much shorter life than the enthusiasm of a group. If your group works from a van or minibus, you can avoid running out of charged batteries by getting an adapter and extension lead in order to use the vehicle's battery. This means you only have to use your batteries when you're not near the van.

Microphone A separate microphone is essential for good sound quality. A uni-directional or cardioid type is the best all-rounder for most groups. The cable needs to be a reasonably long one so that the camera and VCR can be kept a long distance from the person who is talking. To prevent the popping "p's", and to help reduce wind noise, use a foam rubber pop shield. A "fish pole" (or a broom handle with the microphone taped on to it), is another accessory. It allows the operator to stay out of the way whilst keeping the microphone pointed in the right direction.

General Make sure you buy an extension cable and a four-outlet plug board. This is a most useful accessory, since you will have a VCR, television and lights to plug in. Make sure your extension cable is of a good length.

A basic tool kit is also a useful investment. It lets you make repairs when you're at a session. The vast majority of faults in video equipment are caused in the cables and sockets – in the bits that people can touch. It may be useful to carry a soldering iron and solder, a small screwdriver and pliers/wire cutters. Aerosol cans of switch cleaner and video head cleaner, and a roll of sticky tape complete the basic kit.

Appendix 1

Funding

Equipment

Assuming the club or institution you work for does not have enough money to buy equipment, who can you approach? The list below should provide you with a starting point:

- The Council (Education, Leisure & Recreation or Social Services Departments);

- Your Regional Arts Association;

- A Foundation or Charitable Trust interested in work with young people. Consult the *Directory of Grant Making Trusts* published by the Charities Aid Foundation in the reference section of your local library;

- Naturally, you should consider borrowing or hiring equipment (not from a commercial company) if you want to do a pilot project or only need the gear for a short time. There may well be a video or media resource centre locally that you can approach. Ask your Principal Youth Officer, the Media Resource Officer of the local education authority or the Regional Arts Association about such facilities.

It's worth approaching as many bodies as you can because you may get part-funding from a number of sources. Regional Arts Associations are often keen to co-operate with local authorities in jointly funding projects.

Whoever you approach, you may be able to make out a better case if you form a consortium of youth groups in your area and make a plan to share the equipment between yourselves.

Running costs

The sort of bodies listed above are worth approaching for these costs as well. But it's important to go to the right source for the particular work you're doing. For example:

- if you're planning a Playscheme, go to the Leisure and Recreation Department of the Council. Remember to apply early in the year;

- if you're working with young people in trouble with the law or "at risk", approach the Intermediate Treatment Officer in the Social Services Department;

- if you want to pay the labour costs of a video worker most statutory Youth Centres will have tutor hours available on a sessional basis for this sort of work;

- if you're working with a group who feel strongly that they want their idea on broadcast television you should contact Open Space (Community Programme Unit, BBC) or Channel 4.

Appendix 2

The technical bit: How to operate the equipment

This section of the book is not designed as a replacement for the equipment manufacturer's manual. There are so many different models on the market that they cannot all be mentioned here. However, the manual which comes with the equipment does not always stay with it, so the basics are covered here as an aid to memory – a sort of mini-manual. It's compiled as an average for most domestic equipment (which varies considerably) but not for U-Matics.

This mini-manual also recognizes that things can go wrong, and attempts to deal with the most common problems. Having said that, it must be emphasized that there is no substitute for knowing the idiosyncrasies of individual models.

Setting up

The basic kit consists of a camera, a VCR, a monitor/TV, an AC mains tuner/charger, leads (power lead and Co-Ax lead for the VCR and monitor), a tripod, and a microphone and lead. It's also useful to have a camera cable extension (about 5 meters) and a power cable extension, ideally a "4-Way Trailing Socket" (about ten meters long).

Battery power

If you're recording away from mains power, the VCR needs to be portable and to run on batteries. Some receiver/monitors also have a battery or a socket for DC power, such as a car battery.

A day before going out, check that the batteries are fully charged. Check again just before you go out. There is usually a BATTERY CHARGE indicator on the VCR.

Some cameras have an indicator in the viewfinder which blinks on when battery power runs too low. If not, the first sign is usually that either the image in the camera viewfinder begins to go fuzzy or squiggly lines appear on it and the image starts to break up.

You can get adapters which allow portable VCRs and monitors to be run from the cigar lighter socket in a vehicle, or another which attaches directly to the battery. Car batteries will obviously have a much longer life than the smaller batteries which fit directly into equipment.

Battery charging

When you finish shooting, get into the habit of charging the batteries as soon as you return to base, even before you put the kettle on. There's nothing quite as annoying as getting set to go off shooting on location and discovering that someone forgot to charge them up. Leaving batteries in a dead or flat condition for too long will actually cause permanent damage to them – and they're very expensive.

It takes at least four hours, and often as much as eight, to fully re-charge a battery. The charging circuit will cut out automatically, so have no fears about overcharging.

If a battery is inserted in a VCR it will usually charge automatically when the machine is connected to mains power (and switched on), but not when it's recording or playing back.

AC Mains Adapters can also be used for charging. On some the battery attaches directly to the unit; on others it has to be plugged in to a BATTERY or DC OUT socket. There's often an INDICATOR LAMP which is supposed to indicate when the battery is being charged. However, these can have a mind of their own and will often go out long before the battery is fully charged. It's better to rely on a "6 hours for full charge" rule than any INDICATOR LAMP.

Recording from a camera

Set out below are the procedures you will need to follow to record from a camera:

- Set up the monitor and VCR (and AC adapter, if portable);

- Set the camera on its tripod, ensuring that it's not pointed at any bright light source;

- Plug the camera cable into its socket (usually marked CAMERA or CAM) on the VCR. These plugs only connect in a certain way, so check that the notch is located correctly before you push it in. If the camera cable is not the type which is permanently attached to the camera (i.e., it has a plug at each end) it is a good idea to avoid confusion and potential mishaps by marking the camera end with paint or coloured adhesive tape;

- Switch the VCR input selector switch to CAMERA position. (It won't work on TV or LINE);

- Set the TRACKING control on the VCR to its central position;

- Insert the video cassette. An arrow on the cassette indicates the way it goes in. It won't work in any other position so don't apply force without checking. Before you put a cassette in make sure the safety tab has not been broken off (or on U-Matic, that the red button is in place). Like audio cassettes, these tabs can be broken off to prevent a tape being over-recorded;

- It's a good idea to play back a section of the tape to ensure that irreplaceable work is not about to be erased and recorded over;

- If applicable, switch the camera on. Take the lens cap off. Remember: be sure that the camera is never pointed at the sun or a bright light as this can burn the tube permanently and leave black blotches on the picture;

- Press RECORD and PAUSE simultaneously on the VCR. You should have a picture on the monitor (and in the camera viewfinder). If not, check the connections first, then the tuning on the monitor. Presuming that all the above have been covered, and there's still no picture, you need to fall back on the specific manual for your equipment;

- Set the colour temperature on the camera. This is variously marked with symbols for artificial/tungsten light (a light bulb) or natural light (the sun) or, on more advanced cameras, its masked with °K (Kelvin) – in which case 3400 °K is artificial/tungsten and 6000 °K is daylight. An ND marking means neutral density filter which is used for very bright sunlight;

- Set the white balance on the camera by pointing it at a white card (filling the frame) and pressing the white balance button for a few seconds. Some cameras use a translucent white lens cap for this. On others you actually need to set the amount of red and blue (the tube takes care of the other primary colour, green) until the picture looks about right;

- You're now ready to go. Once the RECORD function is pressed on the VCR, the PAUSE switch on the camera will control the VCR. Remember that the tape will be damaged if the VCR is kept on PAUSE for too long. Most VCRs have an automatic cut-out to prevent this by switching the VCR off after a few minutes on PAUSE;

- Always record and play back a short test piece to check that it's actually recording correctly at the beginning of a session.

Troubleshooter

If there's no picture in the camera viewfinder:

- is the camera switched on?

- is the viewfinder plugged into the camera (if it's a separate unit)?

- has the lens cap been removed?

- move the zoom to WIDE ANGLE. When it's on full TELEPHOTO,the lack of focus can confuse the operator into thinking that there is no picture at all;

- check that the aperture is not closed by the MANUAL OVERRIDE control. Under normal circumstances, the aperture would be wide open (f-1.2 to 2.8) indoors, or closed down somewhat outdoors where the light is better (f-5.6 to 16). Sometimes it's set on MANUAL and so will not adjust automatically;

- some cameras have a blackout section on the filter ring which blocks off the iris to protect the tube; this may need to be moved so that the iris lets light through;

- VCRs often have an automatic cut-out circuit for protection against moisture. If there is a lot of moisture in the air, or condensation (often caused by bringing VCRs into warm rooms from cold cars) you can only wait for it to dry out. Warm air, supplied by a fan heater, will help. As a final point, check that the COLOUR MONITOR is correctly adjusted for brightness, contrast and colour balance. Because the MONITOR is used to set up the camera and how a picture looks, it obviously needs to be right. The best way to do this is to use the VCR to record five minutes or so of the Test Card from broadcast TV in the daytime. The relationship between the colours is very precise and enables accurate adjustment of the monitor.

Light

The amount of light that gets into the camera is important. Too much may damage the tube, just as it would if pointed at the sun or a bright light source. Insufficient light produces only a dark and muddy image.

If you're indoors and the aperture is wide open (f-1.4 to f-1.8) but the image is still muddy, greenish and too dark, you need more light. Light comes in only two varieties: natural and artificial. When the sun won't co-operate, turn on all the lights you can. If this is still insufficient, the first step is to replace the lamps with more powerful ones. Several 150 watt bulbs will provide sufficient illumination for basic work. Or you may need to buy or borrow extra lamps.

Shooting outdoors gives another set of problems. The primary one is contrast, or rather the inability of video to cope with contrast. In a situation where there is a large contrast between the brightness of two parts of the picture, a video camera will invariably choose to expose correctly for the brightest part. For example, when shooting people against the skyline, the camera will expose correctly for the sky and leave the people as dark silhouettes. The only way around this is not to shoot them against the sky or to adjust the iris manually to compensate. A darker background would produce a better image altogether.

It's worth pointing out that the camera viewfinder acts only as a rough guide and is often badly adjusted. Viewing the image on a TV or monitor screen is preferable.

When the sun won't co-operate turn on all the lights you can

Sound

The quality of sound on video is very important for the finished programme and not too difficult to obtain. However, standard camera microphones pick up noise from all around, most of which you don't want. Remember: a separate microphone is a better idea. When you plug it into the socket marked MIC on the VCR, it will automatically cut out the camera microphone, so the only sound on the tape is from the better quality separate microphone.

Because it will pick up the noise caused by the movement of your hand against it (as loud thumps and scrapes), you have to hold it firmly and very still. Better yet, use it with a stand to isolate it completely.

Working outside can mean wind noise getting into the microphone. A pop shield will prevent much of it. These inexpensive extras are actually designed to cut down unwanted noise from speech, as much as popping "p's" but they're also very good at decreasing wind noise. In an emergency you can make your own from a piece of foam rubber or even from a sock or folded handkerchief.

Microphone cables are easily pulled from their plug sockets so make a habit of checking it regularly during a shoot. This risk can be reduced by taping the plug to the VCR and the cable to its carrying strap.

The EARPHONE socket on the VCR lets you use an earphone or headphones to check whether or not you're getting sound into the VCR. Most VHS and BETAMAX VCRs have an automatic volume control mechanism so you don't have to adjust the sound levels yourself.

Appendix 3

Resource centres

The North

Newcastle Media Workshop
67-75 Westgate Road
Newcastle-upon-Tyne NE1 1SG
Tel 0632 322410

Northern Arts
10 Osborne Terrace
Newcastle-upon-Tyne NE2 lNZ
Tel 0632 81334

Research Training Initiatives
l8-20 Dean Street
Newcastle-upon-Tyne NE1 1PG
Tel 091 2616581

Swingbridge Video
10a Bridge Street
Gateshead
Tyne & Wear NE8 2BH
Tel 0632 776680

The North East

Blackfriars Arts Centre
Spain Lane
Boston
Lincolnshire PE21 6HP
Tel 0205 63108

Hawthorne Film and Video Workshop
Outreach Community Arts
Northumberland Avenue
Hull HU2 0LN
Tel 0482 226420

Hull Film & Video Workshop
Outreach Community Arts
Northumberland Avenue
Hull HU2 0LN
Tel 0482 226420

Hulltime Based Arts
c/o Salisbury Street
Hull

Leeds Film Group
191 Bellevue Road
Leeds 3
Tel 0532 435653

Lincs & Humberside Arts Association
St. Hughs
23 Newport
Lincoln LN1 3DN
Tel 0522 33555

Postnergate Gallery
6 Postnergate
Hull HU1 2JN
Tel 0482 222745

The North West

Counter Image
19 Whitworth Street West
Manchester MI5 WG
Tel 061 228 3551

Crawford Arts Centre
Mill Lane
Wavetree
Liverpool 15
Tel 051 722 9091

Merseyside Arts Association
6 Blue Coat Chambers
School Lane
Liverpool L1 3BX
Tel 051 709 0671

North West Arts Association
12 Harter Street
Manchester M1 6HY
Tel 061 228 3062

Open Eye Ltd.
90 Whitechapel
Liverpool L1 6EN
Tel 051 709 9460

Workers Film Association
5 Lucy Street
Manchester M15 4BX
Tel 061 848 9783

East Midlands

Burwell House
North Street
Burwell
Cambridge CB5 0BA
Tel 0638 741256

Eastern Arts Association
Cherry Hinton Hall
Cherry Hinton Road
Cambridge CB1 4DW
Tel 0223 215355

East Midlands Arts Association
Mountfields House
Forest Road
Loughborough
Leicester LE11 3HU
Tel 0509 218292

The Minories
74 High Street
Colchester
Essex CO1 1UE
Tel 0206 577067

New Cinema Workshop
c/o Midland Group
24-32 Carlton Street
Nottingham NG1 1NN
Tel 0602 584891

Nottingham Video Project
c/o International Community Centre
61b Mansfield Road
Nottingham NG1 3FF
Tel 0602 470356

33 Film & Video Group
33 Guildford Street
Luton
Bedfordshire LU1 2WQ
Tel 0582 419584

Western Independent Film & Video Association
c/o Magazine Workspace
11 Newarke Street
Leicester LE1 5SS
Tel 0533 559711

London

Actionspace London Events Ltd.
16 Chenies Street
London WC1E 7ET
Tel 01 631 4468

Activision
Unit 20
Mercantile House
All Saints Street
London N1
Tel 01 837 7842

Albany Video
The Albany
Douglas Way
London SE8
Tel 01 692 0231

Anarres
10a Bradbury Street
London N16 8JN
Tel 01 249 9212

Barnet Arts Workshop
Avenue House
East End Road
London N3
Tel 01 346 7120

Black Audio Film Collective
89 Ridley Road
London E8 2NH
Tel 01 254 9536

Ceddo Film & Video Workshop
First Floor
South Tottenham Education & Training Centre
Braemer Road
London N15 5EU
Tel 01 802 9034

CLIO Co-Op
91c Mildmay Road
London N1
Tel 01 249 2551

Fantasy Factory
42 Theobald's Road
London WC1
Tel 01 405 6862

Greater London Arts Association
9 White Lion Street
London N1 9PD
Tel 01 837 88808

Hammersmith & Fulham Video Project
Connections
Palingswick House
241 Kings Street
London W6 9LP
Tel 01 741 1766

InterChange
15 Wilkin Street
London NW5 3NG
Tel 01 267 9421

Inter-Action
Royal Victoria Dock
London E16 1BT
Tel 01 511 0416

Irish Video Project
c/o 118 Talbot Road
London W11
Tel 01 459 1036

Lambeth Video
245a Cold Harbour Lane
London SW9
Tel 01 737 5903

London Video Arts
23 Frith Street
London W1
Tel 01 734 7410

Media Resources Centre
90 Debeauboir Road
London N1 4EN
Tel 01 241 2162

Moonshine
1090 Harrow Road
Kensal Green
London NW10
Tel 01 960 0051

Moonshine Community Arts Workshop
Victor Road
London NW10
Tel 01 969 7959

Retake
25 Bayham Street
London NW1
Tel 01 388 9032

Sankofa
Unit 5
Cockpit Yard
Northington Street
London WC1
Tel 01 831 0024

Star Productions
61 The Basement
Thistlewaite Road
London E5 8QE
Tel 01 986 4470

Tower Hamlets Arts Project
178 Whitechapel Road
London E1 1BJ
Tel 01 247 0216

Ujima Video Workshop
69 Lennox Road
London N4
Tel 01 263 6195/6

West London Media Workshop
118 Talbot Road
London W11 1JR
Tel 01 221 1859

Women In Sync
Unit 6
47 Wharfdale Road
London N1
Tel 01 278 2215

The South

The Media Centre
South Hill Park
Bracknell
Berks RG12 4FA
Tel 0344 427272

The Thamesdown Media Arts Lab
Town Hall Studio
Regents Circus
Swindon SN1 1QS
Tel 0793 26161 ext 3142

Oxford Film Makers Workshop
The Stables
North Place
Headington
Oxford OX3 9HY
Tel 0865 60074

Portsmouth Film and Video Workshop
Portsmouth College of Arts Design and Further Education
Winston Churchill Avenue
Portsmouth
Hampshire PO1 2DJ
Tel 0705 826435

Portsmouth Film Makers Co-operative
Hornpipe
143 Kingston Road
Portsmouth PO2 7ED
Tel 0705 861851

Southern Arts Association
19 Southgate Street
Winchester
Hants SO23 9DQ
Tel 0962 55099

South East Arts
10 Mount Ephraim
Tunbridge Wells
Kent TN4 8AS
Tel 0892 41666

Watershed Media Centre
1 Canons Road
Bristol BS1 5TX
Tel 0272 276444

Wales

Aberystwyth Media Group
The Barn Centre
Alexandra Road
Aberystwyth
Dyfed
Tel 0970 4001

Caernavon Video Project
12 Palace Street
Caernarvon
Cwynedd
Tel 0286 4343

Chapter Video Workshop
Market Road
Canton Cardiff
Tel 0222 396061

North Wales Association
for the Arts
10 Wellfield House
Bangor
Gwynedd LL57 1ER
Tel 0248 353248

South East Wales Arts
Association
Victoria Street
Cwmbran
Gwent NP4 3YP
Tel 06333 75075

Valley & Vale Community Arts
The Old Junior School
Maesteg Road Tongu
Mid Glamorgan
Tel 0656 724799

Welsh Arts Council
Holst House
9 Museum Place
Cardiff CF1 3NX
Tel 0222 394711

West Wales Association
for the Arts
3 Red Street
Carmarthen
Dyfed
Tel 0267 234248

Wrexham Community Video
Church House
Rhossdu Road
Wrexham
Clwyd
Tel 0978 364611

Scotland

Commedia
11 Douglas Terrace
Haymarket
Edinburgh

Film Workshop Trust
17 Great King Street
Edinburgh
Tel 031 556 2078

Glasgow Arts Centre
12 Washington Street
Glasgow G3 8AZ
Tel 041 221 4526

Scottish Arts Council
19 Charlotte Square
Edinburgh EH2 4DF
Tel 031 226 6051

Second City Workshop
17 Kennedar Drive
Glasgow G51

Video In Pilton
Craig Royston High School
Pennywell Road
Edinburgh
Tel 031 332 7801

Working with Girls

If you want to know more about working with girls, you should contact Val Carpenter, Projects Officer at the National Association of Youth Clubs. She can give you more information, or put you in touch with others in your area. They have a resource bank at NAYC for work with girls, including a library of suitable video tapes. The Association can be reached on 0533 29514. The address is: Keswick House, 30 Peacock Lane, Leicester LE1 5NY.

Appendix 4

Video distributors

Albany Video Distribution
The Albany
Douglas Way
London SE8 4AG
Tel 01 692 6322

Cinema of Women
Unit 313
31 Clerkenwell Close
London EC1R 0AT
Tel 01 251 4978

Circles
113 Roman Road
London E2 0HU
Tel 01 981 6828

Concord Films Council
201 Felixtowe Road
Ipswich 1P3 9BJ
Tel 0473 76012

London Video Arts
23 Frith Street
London W1
Tel 01 734 7410

The Other Cinema
79 Wardour Street
London W1V 3TH
Tel 01 734 8508

Team Video
105 Canalot
Kensal Road
London W10
Tel 01 960 5536

The Good Video Guide
ICA
12 Carlton House Terrace
London SW1Y 5AH

Appendix 5

Further reading

There are, of course, a large number of guide-books on the market. Not all of them are very good; others are useful, depending on your needs and interests. The following is a very selective list of introductory books which will get you started.

Focal Guide to Moviemaking by Paul Petzold. Focal Press, £5.50.
Intelligent technical instructions on shooting, editing and displaying film.

All-In-One Cine Book by Paul Petzold. Focal Press, £5.25.
Advice for the beginner on technique and style.

Home Video by Richard Dean. Butterworth, £5.95.
A basic introduction to using video.

The Complete Video Handbook by Owen and Dunton. Penguin, £5.95.
A thorough guide to all aspects of video.

The following video magazines also carry regular features on using video: *Which Video?; Video Magazine.* Flick through them for ideas.

Another useful publication is *Independent Video* which comes out monthly (price 50p) available from The Media Centre, South Hill Park, Bracknell, Berks (Tel 0344 427272).

Glossary

Aperture See Iris

AGC/ALC Automatic Level Control. An electronic circuit that adjusts the incoming signal to a predetermined level. Can apply to either audio or video signals.

Audio dub To re-record the audio portion of a video tape without disturbing the video portion of the signal.

Betamax A half inch home video format.

Barn doors Adjustable flaps attached to the sides of a lamp housing which are used to control the distribution of the light beam being cast onto an area.

BNC A video connector.

Boom A long-armed device used to suspend a microphone, a light or a camera above the performers and out of camera range of the scene being shot.

CCTV Closed Circuit Television. Any form of television that is locally originated and displayed. Non-network and non-cable TV, commonly applied in schools, offices and the like.

Close-up A camera shot in which the main subject of a scene is shown large on the screen with little visible background.

Coaxial cable A specially designed one-ground, one-conductor cable that is shielded and is used to transmit television signals.

Colour temperature The relative degree of colour quality (reddish, yellowish, bluish, whitish) produced by a given light source. Colour temperature is measured in "degrees Kelvin" and is important for matching the video camera's characteristics with the lighting conditions.

Control track The portion along the length of a video tape on which information is placed and used to control the recording and playback of the video signal on a video recorder.

Crash edit An extremely simplified method for editing, which uses two ordinary VCRs.

Cardioid One of the many possible pick-up patterns of microphones. This particular heart-shaped pattern is most sensitive to sounds coming the front of the microphone, less so from the side and rejecting sounds coming from the back.

Depth of field The distance between the nearest object in focus and the farthest object in focus.

Din plugs Can be three-, five- or six-pin plugs, depending on their use, although they all have the same outer appearance.

Dubbing Electronic duplicating of recorded video and/or audio information.

Dynamic microphone A very pressure-sensitive microphone of a moving-coil design. This type of microphone can be either uni- or omni-directional and can stand up to rough handling.

Edit To re-record the original video material in a sequence that's suitable for presentation.

Editing deck A specially constructed video cassette-recorder which has, in addition to play and record circuitry, circuitry and controls which, in conjunction with another VCR, make very accurate edits.

Electret condenser microphone A capacitor microphone that is very sensitive and requires a battery power supply.

Fluid head tripod A tripod where the camera rests on two metal plates. The upper rotating plate rests on a bed of fluid and the movement provided is very smooth.

F-stop A control (f4, f5.6, f8, etc.) which indicates the amount of light passing through the lens of the video camera. It can be adjusted to vary the size of the aperture. The higher the f-stop number, the smaller the aperture opening.

Fade (Fade-in/Fade-out) A gradual increase or decrease of either the audio or video signal strength. As a video technique, a fade-in is to bring the image slowly into view out of black; while a fade-out is this technique in reverse.

Framing Both positioning the camera and adjusting its lens in order to include (as well as exclude) all the visual elements required for shooting a particular subject.

Friction head tripod A tripod whose camera mount consists of two metal plates, the lower stationary, the upper rotating; generally does not provide smooth camera movement though more expensive models utilize ball bearings to offset this problem.

Generation The number of re-recordings away from the original. The 1st generation is a recording of the original, the 2nd generation is a recording made of a 1st generation recording, etc.

Iris An adjustable device covering the aperture of the lens, often composed of a set of metal leaves and used to control the amount of light passing through the lens. Iris openings are measured in f-stops.

Lavalier microphone A small microphone designed to be worn either around the neck or clipped to a tie, collar or the like.

Lens A series of optical elements (or the elements themselves), contained within a barrel or tube, which collect and focus light. Employed to project an image onto the target area of the pickup tube in the video camera.

Lux The term applied to the metric measurement of light intensity, taken at the surface on which the light source is illuminating.

Macro lens A lens capable of focusing in on very small minute objects for ultra close-up shots.

Monitor/Receiver See TV monitor.

Moving coil microphone See dynamic microphone.

Omni-directional microphone A microphone design with a pickup pattern which is equally sensitive to sounds coming from any direction.

Pan The horizontal pivoting of the camera, either right or left, during shooting.

Phone plug A variety of plug most often used with audio amplifiers. Also known as an RCA plug.

Pop filter A sponge rubber or plastic foam cap placed over the end of a microphone to reduce breathing sounds, popping "p's" and "b's" and other unwanted vocal effects.

RF converter An electric device which changes the recorded video signals on the video tape into RF (radio frequency) signals, enabling them to be played back over a standard TV set.

Storyboard A script-like presentation with sketches and camera instructions that illustrate the sequences to be shot.

Saticon A colour vidicon-like pickup tube developed by Hitachi and NHK (Japanese Broadcasting Corporation) which has improved characteristics over conventional vidicons.

Telephoto lens A specially designed camera lens with a longer-than-normal focal length; thus a narrower field of view. Useful for focusing in on subjects and scenes at a great distance from the camera.

Tracking The angle and speed at which the tape passes the video heads. The angle and speed of recording must correspond to that of playback or distortion will result.

Tracking shot Following the action by moving the camera itself along with the movement of the subject being shot.

U-Matic The three quarter inch format video cassette-recorders and players in which the tape is wrapped about the head drum in a U shaped pattern.

Uni-directional microphone A microphone which picks up sound from one direction only.

VCR Video cassette-recorder.

VHS Video Home System. The half inch home video format.

Video cassette-recorders Portable VCRs, especially those for VHS or Betamax, are often sold along with a particular camera. Some cameras and VCRs are not inter-changeable without different plugs.

Videodisc An electro-mechanical system that uses a solid "record" from which recorded video and audio information can be played back and viewed over a standard TV receiver.

Viewfinder An electronic or optical device through which the video camera operator can view the scene being shot. Electronic viewfinders generally permit viewing the playback of recorded material.

White balance The camera adjustment which sets the colour video camera to the proper white reference.

Wide angle lens A special type of camera lens design having a very short focal length and a very extended horizontal field of view. Used for shooting broad, panoramic scenes.

Zoom lens A camera lens with a variable focal length that allows a continuous changing of the field of view from wide angle to telephoto or vice versa.

About...

Tony Dowmunt has been using video with young people for the last twelve years. He has worked with young people in a wide variety of ways and places – schools, youth clubs, playschemes and small groups. He was a founder member of the Walworth and Aylesbury Community Arts Trust, set up in 1975. For the last seven years he has been with Albany Video, a community based workshop in Deptford, London. He is currently setting up an independent video production company. He lives in Islington with Maggie Mckenzie and Zosia, their daughter.

Judy Farrar has illustrated several books including another in this series entitled *Sounds Good*. She designs for community and fringe theatre companies and has made many sets and props for the Spare Tyre Theatre Company. Recently, she has finished designing the theatre company's book, *The Spare Tyre Songbook*. She is presently studying Art Therapy at Goldsmiths College and lives in Hackney with Stan, her six year old son.

Inter-Action is a group of associated not-for-profit Companies and Trusts which follow an approach called Social Enterprise – the application of entrepreneurial skills to obtain direct social and community benefits.

Inter-Action uses this approach to provide training, consultancy and resources to help others in the voluntary, statutory and small business sectors. The objective of this is to improve their management, programme effectiveness, earnings, fund-raising, cost-savings and their use of under-used facilities.

Inter-Action also develops new model projects and techniques which can easily be used by local groups elsewhere in a cost-effective and self-helping way. Some of the projects which Inter-Action has helped pioneer throughout the U.K. and other European Countries are City Farms, Community Computer Camps, mobile play facilities, community architecture and technical advice agencies, community resource centres and printshops.

Community Media Division of Inter-Action under the direction of Peter Mount, provides all the technical productions and resources for the charity as well as training for other agencies and model project development work. This includes tape-slide shows, video production, a darkroom, print and publishing services, community radio and recording facilities and an embryonic interactive video-disc capability.

The Division helps young bands produce demo-tapes and helps other agencies set up Breaks for Young Bands centres to benefit hundreds of young bands in their own locale. It runs a Community Media Van, a media studio on wheels, which enables anyone, anywhere – from playschemes to street markets – to participate in making community, video and radio programmes with accompanying local news-sheets.

Inter-Action In-Print is the publishing wing of Inter-Action Trust. It has recently produced the Inter-Action Handbooks of which this book is an example. Other Series titles include *Sounds Good, Community Computer Projects* and *Which Software? Which Hardware?* Other books in print include *Breaks For Young Bands, Prof. Dogg's Game Songs* and *Make A Real Job Of It.*

British Library Cataloguing in Publication Data:

Dowmunt, Tony
Video with young people.—(Inter-Action Handbooks).
1. Video tapes in social work with youth —Great Britain 2. Video tapes in social group work—Great Britain
I. Title
362.7 HV1441.G7

ISBN 0 304 31486 2